Successful Management in Foodservice Operations

David K. Hayes
Jack D. Ninemeier

SENIOR EDITORIAL DIRECTOR	Justin Jeffryes
EXECUTIVE EDITOR	Todd Green
EDITORIAL ASSISTANT	Kelly Gomez
SENIOR MANAGING EDITOR	Judy Howarth
PRODUCTION EDITOR	Mahalakshmi Babu
COVER PHOTO CREDIT	© andresr/Getty Images

This book was set in 9.5/12.5pt STIX Two Text by Straive™.

Published by John Wiley & Sons, Inc., Hoboken, New Jersey.
Published simultaneously in Canada.

This book is printed on acid-free paper.

Founded in 1807, John Wiley & Sons, Inc. has been a valued source of knowledge and understanding for more than 200 years, helping people around the world meet their needs and fulfil their aspirations. Our company is built on a foundation of principles that include responsibility to the communities we serve and where we live and work. In 2008, we launched a Corporate Citizenship Initiative, a global effort to address the environmental, social, economic, and ethical challenges we face in our business. Among the issues we are addressing are carbon impact, paper specifications and procurement, ethical conduct within our business and among our vendors, and community and charitable support. For more information, please visit our website: www.wiley.com/go/citizenship

Library of Congress Cataloging-in-Publication Data applied for

ISBN: 978-1-394-20849-4 (PBK)

The inside back cover will contain printing identification and country of origin if omitted from this page. In addition, if the ISBN on the back cover differs from the ISBN on this page, the one on the back cover is correct.

SKY10076268_053124

Contents

Preface

Those who own or manage a foodservice operation want it to be successful.

The definition of what constitutes a successful foodservice operation, however, varies widely. In some cases, success is measured by the number of guests served or the total amount of revenue an operation generates.

For most commercial (for-profit) operations, success may be measured by both profits produced and an owner's return on their investment. In many other settings, a successful operation may mean one with minimized operating costs or one that optimizes the nutritional value of the meals it serves.

Regardless of the definition of success, those operating successful foodservice facilities must be very talented individuals. Unlike many other managers, foodservice operators must assume the professional role of both a manufacturer and a retailer. A foodservice operation is unique because all the functions of a product's sale from menu development (what items should be offered?) to guest service are often in the hands of the same individual.

Foodservice operators are responsible for securing raw materials, producing products, and selling them—all under the same roof. Few other managers are required to have the breadth of skills that effective foodservice operators must have.

While there are different ways to define success in a foodservice operation, the ability to operate a successful foodservice business requires:

✓ Serving high-quality products
✓ Providing high-quality service
✓ Generating optimal levels of operating profits

The purpose of this book is to teach foodservice operators what they must know, and do, to achieve their own definitions of success. This book is distinctive in that it addresses those success factors that are unique to the foodservice business. Among the many key topics addressed in this book are:

✓ The unit operator's role in successful foodservice management
✓ How to read a Uniform Systems of Accounts for Restaurants (USAR) income statement

✓ How to calculate an operation's break-even point
✓ How to identify a target market
✓ How to create a marketing plan
✓ How to create an effective proprietary website
✓ How to manage marketing on third-party operated websites
✓ How to create and manage a profitable menu
✓ How to create and manage off-premise (takeaway) menus
✓ How to successfully price menu items
✓ How to recruit and select effective team members
✓ How to train new team members
✓ How to control food and beverage production costs
✓ How to control service delivery costs
✓ How to manage labor costs
✓ How to develop a revenue security program
✓ How to prepare an accurate operating budget
✓ How to monitor and modify an operating budget

Most importantly, all of these topics and more are addressed in ways specific to foodservice operations of all sizes and in all industry segments.

Readers will quickly find that the content of this book is essential to the successful management of their own operations, and they will also find that the information in each chapter has been carefully selected to be easy-to-read, easy-to-understand, and easy-to-apply.

Book Features

In addition to the essential employee management information it contains, special features were carefully crafted to make this learning tool powerful but still easy to use. These features are:

1) **What You Will Learn.** To begin each chapter, this very short conceptual listing summarizes key concepts readers will know and understand when they complete the chapter.
2) **Operator's Brief.** This chapter opening overview states what information will be addressed in the chapter and why it is important. This element provides readers with a broad summation of all important issues addressed in the chapter.
3) **Chapter Outline** This two-level outline feature makes it quick and easy for readers to find needed information within the body of the chapter.

4) **Key Terms** Professionals in the foodservice industry often use very special terms with very specific meanings. This feature defines over 200 important (key) terms so readers will understand and be able to speak a common language as they discuss issues with their colleagues in the foodservice industry. These key terms are defined when they are first used in a chapter, are listed at the end of each chapter in the order in which they initially appeared and are listed in an alphabetized glossary at the book's end.

5) **Find Out More** In a number of key areas readers may want to know more detailed information about a specific topic or issue. This useful book feature gives readers specific instructions on how to conduct an Internet search to access that information and why it will be of importance to them.

6) **Technology at Work** Advancements in technology play an increasingly important role in many aspects of foodservice operations. This feature was developed to direct readers to specific technology-related Internet sites that will allow them to see how advancements in technology can assist them in reaching their operating goals.

7) **"What Would You Do?"** These "mini" case studies, located in every chapter of the book, take the information presented in the chapter and use it to create a true-to-life foodservice industry scenario. They then ask the reader to think about their own response to that scenario (i.e. *What Would You Do?*).

This element was developed to help heighten a reader's interest and to plainly demonstrate how the information presented in the book relates directly to the practical situations and challenges foodservice operators face in their daily activities.

8) **Operator's 10-Point Tactics for Success Checklist** Each chapter concludes with a checklist of tactics that can be undertaken by readers to improve their operations and/or personal knowledge. For example, in a chapter of the book addressing "Accounting and Financial Management in Foodservice Operations" one point in that chapter's 10-Point Tactics for Success Checklist is:

> *Operator understands the purpose of the Uniform System of Accounts for Restaurants (USAR) and the advantages of using it in the preparation of financial reports.*

Instructional Resources

This book has been developed to include learning resources for instructors and for students.

To Instructors

To help instructors (and corporate trainers!) effectively manage their time and enhance student-learning opportunities, the following resources are available on the instructor companion website at www.wiley.com/go/hayes/managementfoodservice.

✓ Instructor's Manual that includes author commentary for "What Would You Do?" mini case-study questions.
✓ PowerPoint slides for instructional use in emphasizing key concepts within each chapter.
✓ A 100-item Test Bank consisting of multiple-choice exam questions, their answers, and the location within the book from which the question was obtained. The test bank is available as a print document and as a Respondus computerized test bank. Note: **Respondus** is an easy-to-use software program for creating and managing exams that can be printed to paper or published directly to Blackboard, WebCT, Desire2Learn, eCollege, ANGEL, and other e-learning systems.

To Students

Learning about the successful management of a foodservice operation will be fun. That's a promise from the authors to you. It is an easy promise to make and keep because working and managing in the foodservice industry is fun. And it is challenging. However, if you work hard and do your best, you will find that you can master all of the important information in this book.

When you do, you will have gained invaluable knowledge that will enhance your skills and help advance your hospitality career. To help you learn the information, for this book, online access to over 225 Power Point slides is available to you. These easy-to-read tools are excellent study aids and can help you when taking notes in class.

Acknowledgments

Successful Management in Foodservice Operations has been designed to be the most up-to-date, comprehensive, technically accurate, and reader-friendly learning tool available to those who want to know how to effectively management a foodservice operation of any type.

The authors thank Catriona King of Wiley for initially working with us to develop the idea for a series of practical books that would help foodservice operators of all sizes more effectively manage their businesses. She was essential in helping conceptualize the need for this book as well as all of the other books in this five-book Foodservice Operations: essential series. The five titles in the *"Foodservice Operations: Essential"* series are:

✓ *Successful Management in Foodservice Operations*
✓ *Marketing in Foodservice Operations*
✓ *Accounting and Financial Management in Foodservice Operations*
✓ *Cost Control in Foodservice Operations*
✓ *Managing Employees in Foodservice Operations*

We would also like to thank the external reviewers who gave so freely of their time as they provided critical industry and academic input on this series. To our reviewers, Dr. Lea Dopson, Gene Monteagudo, Isabelle Elias, and Peggy Richards Hayes we are most grateful for your comments, guidance, and insight. Also, thanks to Biloxi Mississippi's Michael T. Kavanagh, who was a technological friend indeed, when we were most in need!

Books such as this one require the efforts of many talented specialists in the publishing field. The authors were extremely fortunate to have Todd Green, Judy Howarth, and Kelly Gomez at Wiley as our publication team. Their efforts and creativity went far in helping the authors present the book's material in the best and clearest possible form.

Finally, the authors would like to thank the many students and industry professionals with whom we have interacted over the years. We sincerely hope this book allows us to give back to them as much as they have given to us.

David K. Hayes, Ph.D.
Jack D. Ninemeier, Ph.D.

Dedication

The authors are honored to have the opportunity to dedicate this book, and the entire Foodservice Operations: Essentials series, to two outstanding and unique individuals.

Brother Herman Zaccarelli

Brother Herman E. Zaccarelli, C.S.C., passed away in 2022 at the Holy Cross House in Notre Dame, Indiana. His professional work included many projects for the hospitality industry, and he published several books and hundreds of articles for numerous trade publications over many years. Among numerous accomplishments, Herman founded Purdue University's Restaurant, Hotel, and Institutional Management Institute in 1976. Later, he served as Director of Business and Entrepreneurial Management at St. Mary's University in Winona, Minnesota.

A lifelong learner, at the age of 68, Brother Herman retired to Florida where he earned a Bachelor's degree in Educational Administration and a Master's degree in Institutional Management.

Herman's ideas and concepts have been widely adopted in the hospitality industry, and he assisted many young educators including the authors of this book series. He will be remembered as a colleague with creative ideas who provided significant assistance to those studying and managing in the hospitality industry. Herman was especially helpful in discovering and addressing learning opportunities for Spanish speaking students, educators, and managers throughout the United States and around the world.

Dr. Lea R. Dopson

A lifelong friend, advisor, and colleague, as well as an outstanding author herself; at the time of her untimely passing, Lea served as President of the International Council on Hotel, Restaurant, and Institutional Education (ICHRIE) and Dean of the prestigious Collins College of Hospitality Management (Cal Poly Pomona).

Lea was a dedicated hospitality professional and a fierce advocate for hospitality students at all levels. Those who knew her were continually in awe of her intelligence and humility.

It was especially fitting that Lea was named as a recipient of the H.B. Meek Award. That award is named after the individual who started the very first hospitality program in the United States (at Cornell University). Selected by the recipient's peers, it goes not to the most outstanding academic professional working in the United States but to the most outstanding academic professional in the entire world. That was Lea.

While she is dearly missed, her inspiration goes on everlastingly in the works of the authors.

1

The Operator's Role in Successful Foodservice Management

What You Will Learn

1) Keys to Success in Foodservice Operations
2) Foodservice Operators' Essential Tasks
3) The Importance of Serving Food Safely
4) The Importance of Serving Nutritious Food

Operator's Brief

In this chapter, you will learn that those operating successful foodservice businesses are unique individuals. Unlike most other managers, they are responsible both for producing products and selling them to the end user (guests).

While success in a foodservice operation can be defined in various ways, the success of every foodservice operation depends on three key factors:

1) Serving high-quality products
2) Providing high-quality service
3) Generating optimal levels of operating profits

Serving high-quality products is not the same as selling expensive products. All foodservice operations, regardless of their menu prices, must provide their guests with high-quality products if they want those guests to return again and again.

While foodservice guests seek high-quality products, they also seek high-quality service. Whether an operation serves its guests from the window of a food truck or in the most luxurious of dining rooms, service levels must be

(Continued)

high. In this chapter, you will learn about factors that influence a foodservice operator's ability to provide guests with high-quality service. You will learn that both commercial (for-profit) and noncommercial (not-for-profit) foodservice operators must carefully manage their revenue and expenses to achieve their financial goals.

Foodservice operators must complete essential tasks if they are to be successful. These tasks include:

✓ Planning
✓ Organizing
✓ Directing and Leading
✓ Controlling
✓ Evaluating

In this chapter, you will learn why each of these tasks is critical.

Finally, you will learn that foodservice operators must provide guests with foods that are safe to eat and that are nutritious. Food safety is critical to an operation's success, and every operation is subject to regular governmental inspections to help ensure that menu items they sell are prepared and handled safely.

The nutritional value of food is important both for the health of guests and for an operation's ability to attract and retain them. Guests are increasingly concerned about the nutritional aspects of the foods they buy, and, in this chapter, you will learn what back-of-house and front-of-house staff must know and do to address these concerns.

CHAPTER OUTLINE

Keys to Success in Foodservice Operations
    Serving High-quality Products
    Providing High-quality Service
    Generating Optimal Levels of Operating Profits
Foodservice Operators' Essential Tasks
    Planning
    Organizing
    Directing and Leading
    Controlling
    Evaluating
Serving Food Safely
Serving Nutritious Food
    Nutrition Essentials for Back-of-House Staff
    Nutrition Essentials for Front-of-House Staff

Keys to Success in Foodservice Operations

Those who own or manage a foodservice operation want it to be successful. The definition of what constitutes a successful foodservice operation, however, varies widely. In some cases, success is measured by the number of guests served or the total amount of revenue an operation generates. For most commercial (for-profit) operations, success may be measured by both profits produced and an owner's return on their investment. In many other settings, a successful operation may mean one with minimized operating costs or one that optimizes the nutritional value of the meals it serves.

Regardless of the definition of success, those operating successful foodservice facilities must be very talented individuals. Unlike many other managers, foodservice operators must assume the professional role of both a manufacturer and a retailer. A foodservice operation is unique because all the functions of a product's sale from menu development (what items should be offered?) to guest service are often in the hands of the same individual.

Foodservice operators are responsible for securing raw materials, producing products, and selling them—all under the same roof. Few other managers are required to have the breadth of skills that effective foodservice operators must have. Since foodservice operators are in the service sector of business, many aspects of management are even more challenging for them than for their manufacturing or retailing management counterparts. The reason: a foodservice operator is one of the few types of managers who have contact with the ultimate consumer (their guests).

This is not true, for example, with the manager of a cell phone factory or an automobile production line. These individuals produce a product, but they do not sell it to the person who will actually use it. In a like manner, furniture or clothing store operators sell products to those who use them, but they normally have no role in producing the products they sell.

The face-to-face guest contact in the foodservice industry requires operators to assume the responsibility of standing behind their own work and the work of their staff, in a one-on-one situation with the ultimate consumer (the guests who purchase the products and services!). Doing so requires focused knowledge and skill, and this book provides an overview of the knowledge and skill areas required to successfully operate a foodservice business.

While there are different ways to define success in a foodservice operation, the ability to operate a successful foodservice business requires:

1) Serving high-quality products
2) Providing high-quality service
3) Generating optimal levels of operating profits

Serving High-quality Products

Successful foodservice operations must sell high-quality products. This is true whether the foodservice operation's guests pay high or modest prices for the products they purchase. Regardless of what they sell, foodservice operators face four very specific product-specific challenges. These challenges relate to:

✓ Quality
✓ Quantity
✓ Delivery method
✓ Value perception

Quality

Every foodservice operator's menu items must be made with wholesome food that is safe to eat. However, the quality of ingredients used to make menu items vary greatly, and meats, fruits, cheeses, coffee, wine, and other alcoholic beverage products provide just a few examples.

For example, the New York strip steak served in one steakhouse may be classified as USDA Prime, the highest quality ranking for beef that has been graded by the United States Department of Agriculture (USDA). A competitor of that operation may serve New York strip steaks graded as USDA Choice, the second highest quality ranking. In this example, both operations may advertise that they sell New York strip steaks, but the quality of the steaks is not the same, nor is the purchase price the operator must pay to a meat supplier for the steaks.

As a result, it becomes a challenge for the operation serving the higher quality Prime steaks to indicate that higher menu prices must be charged because guests are served the highest possible quality of beef. Whether the menu items to be served are burgers, pizzas, or Prime steaks, the quality level of a foodservice operation's menu items must be consistent, and quality must be effectively communicated to guests.

Quantity

There are no legal requirements dictating the quantity of a menu item that must be served in a foodservice operation. A food operator might choose to serve a 3-, 4-, 5-, or 6-ounce hamburger patty. Similarly, an operation's "Fried Fish Basket" may include 1, 2, or 3 fried fish fillets of varying sizes. Soft drinks may be sold as small, medium, or large, with the drink's container size determined by the operator.

Since portion sizes vary, new guests frequently have little idea about the actual quantity of food they are buying when they place a food order. Detailed menu descriptions and photographs of food items can help. However, guests must be alerted to the quantity of food or beverage they will receive for the amount they will pay to be consistently satisfied with their purchases.

Delivery Method

Different foodservice operators offer identical products in identical portion sizes and yet, from their guest's perspectives, the products received are very different. For example, one operator may serve a 5-ounce cup of coffee made with a particular brand of ground coffee and served in a paper cup with a plastic lid and provided to guests as they go through the operation's drive-thru window.

Another operator may offer the same 5-ounce cup of coffee made with the same brand of coffee. However, it is served on a saucer in an elegant china cup in an exclusive up-scale dining room. In this example, both product quality and quantity are identical, yet due to the menu item's delivery, what each operation's guests receives is very different.

In addition to the menu items they desire, some guests will want their menu selections served in a particular way perhaps in a particular container, in a specific setting, or at a particular level of speed. In many locations, guests have numerous alternatives including **quick service restaurants (QSRs)**, fast casual restaurants, or fine-dining operations. When guests have these alternatives available, it can be a challenge to communicate to them how an operator's product offerings are differentiated from those of other operators.

Value Perception

All foodservice operators must provide their guests with good **value** for the products purchased. If they do not, guests are unlikely to be satisfied and will not return. The challenge for foodservice operators relative to value perception is to remember that, in any business transaction, value is determined by the buyer not by the seller.

For example, a foodservice operator may sincerely believe that a menu item's quality, quantity, and delivery method will provide good value to guests if it is sold for $19.95. If, however, too few customers share that view of value, the operation will not consistently be able to sell the menu item.

Key Term

Quick service restaurants (QSRs): Foodservice operations that typically have limited menus that often include a counter at which customers can order and pick up their food. Most QSRs also have one or more drive-thru lanes that allow customers to purchase menu items without leaving their vehicles. QSRs may also offer off-site delivery services for their menu items. Menu prices in QSRs are normally lower when compared to some other restaurant types.

Key Term

Value: The amount paid for a product or service compared to the buyer's view of what they receive in return.

Foodservice operators must also understand that providing good value to guests is not the same as providing low-selling prices. Guests may perceive excellent levels of value in both high-priced foodservice operations and in those that charge more modest prices. Alternatively, guests may pay very low prices for menu items and still feel they did not receive good value for the money they spent.

Experienced operators know their customers assign the value of many foodservice products they purchase based on factors in addition to quality, quantity, and method of delivery. This means that communicating the true value of items sold can be extremely challenging.

To illustrate how the concept of value actually works, consider that, when service levels are identical, if a customer normally pays $19.99 for a medium pizza, a larger pizza for the *same* price represents an increase in value. A smaller pizza for the same price represents a *decrease* in value. Similarly, a reduction in price from $19.99 to $16.99 for a medium pizza represents an increased value.

In most cases, buyers make their value judgments regarding the wisdom of a purchase based on their own personal assessment value. As a result, an important aspect of marketing foodservice products entails communicating to guests exactly "what they will get" and "why it is of value to them." This becomes even more challenging because the foodservice industry, as part of the hospitality industry, is classified as a **service industry**, rather than a manufacturing industry.

Key Term

Service industry: A business segment that primarily provides services to its customers.

Foodservice operations prepare (manufacture) food and beverage items that are then served to their guests. Clearly, the steaks and glasses of wine served to guests in a fine-dining restaurant are products. The presentation of the steaks and wine, however, is clearly a service. Perhaps one reason foodservice operations are classified as members of the service industry relates to the importance of product *and* service quality to the guests' perceptions of value. Also, it is true that the most successful foodservice operators understand that most guests perceive service quality to be *more* critical to the assessment of value delivered than is product quality.

Providing High-quality Service

Foodservice operators sell products, but they deliver services. In most cases, it is easier for foodservice operators to control the quality of the products they sell than to control the quality of the service levels they provide.

For example, a 16-ounce rib eye steak will always be larger than a 10-ounce rib eye. Priced properly, guests may perceive better value in the larger steak (such as when the selling cost per ounce is less for the larger steak). Similarly, French fries may be offered in small or large portion sizes. In both examples, the buyer can easily be told what they will be receiving for the prices they will be charged.

When a foodservice operation advertises that it offers quality service, good service, or quick service, however, these concepts may not be so easy to communicate and consistently deliver. Consider, for example, that an operation's quality is affected by unique characteristics of service that include:

✓ Intangibility
✓ Inseparability
✓ Consistency
✓ Limited capacity

Intangibility

Foodservice customers buy products, and they receive services. The sale of a service provides an intangible benefit to a foodservice guest. It is intangible because, unlike a physical product, a service cannot be seen, tasted, felt, heard, or smelled before its purchase and delivery. In most cases, a service is actually a performance rather than a product. Since services are purchased in an intangible form, operators face unique challenges in communicating the benefits of services offered to those who will buy them.

Guests buying from a foodservice operation must put their faith in the operator that the service provided will be of high quality. Foodservice operators must justify this trust by consistently delivering service at a level that meets and even exceeds their guests' expectations.

Inseparability

Inseparability refers to the tendency of foodservice guests to equate the quality of service they receive with the actual person who provides it. As a result, a *rude* waitstaff member will likely be perceived by guests as providing poor service, while the *cheerful* waitstaff member will be perceived as providing good service. This is true even when the precise tasks performed by the two employees are similar or even identical.

This is the reason that so many foodservice operators hire those who will directly serve guests for their attitudes, rather than for their skills. These operators understand that the real job of a professional food server is to:

✓ Make guests feel welcome
✓ Make guests feel important
✓ Make guests feel special
✓ Make guests feel comfortable
✓ Show a genuine interest when guest expectations have not been met and corrective actions must be taken
✓ Correct any service shortcomings promptly and with a positive attitude

Consistency

Consistency in service can create challenges because service quality often depends on the individual who supplies it. Inconsistency when providing services is usually much greater than when providing products. For example, the customer at a sports bar will usually find that the quality of bottled beers of the same brand purchased while watching a game is identical. However, the skill level, appearance, amount of attention provided, and attitude of the beer's server can vary greatly, even during the same game.

In this example, potential differences in product delivery can have a direct effect on the quality of the customer's beer purchase experience. Due to the importance of consistency when providing hospitality goods and services, it follows that foodservice operators should pay a great deal of attention to their operation's **front-of-house staff** training and standardization efforts.

Limited Capacity

Regardless of their type, nearly all foodservice operations face limitations in their capacity. In front of the house (public areas), capacity may be limited by the number of seats in the dining room or the number of drive-thru lanes and windows available.

In the **back-of-house staff** areas (employee-only areas), capacity may be limited by the number of workers available or the production capability of an operation's preparation equipment. As a result, it may be easier to deliver high-quality service during times of lower ("average") capacity than when at full capacity. However, foodservice operators must, whenever possible, operate at full capacity because, unlike product inventory that can be held over from one day to the next, lost revenue from a seat remaining empty during a specific service period can never again be generated.

Key Term

Front-of-house staff: The employees of a foodservice operation whose duties routinely put them in direct contact with guests.

Key Term

Back-of-house staff: The employees of a foodservice operation whose duties do not routinely put them in direct contact with guests.

Since most foodservice operations face limited capacity, it is important that staff are assigned to work based on anticipated guest demand levels. It is true that many workers scheduled during slow times may cause an operation's labor-related costs to rise to unacceptable levels. However, too few workers available during busy times may cause an operation's service levels to suffer, even when service workers are well-trained and doing their best to assist guests.

Technology at Work

Experienced foodservice professionals know that server training for their employees is essential for ensuring guest satisfaction and for improving the professionalism of their operations' guest service teams.

In some foodservice operations, the amount of server training required may be extensive. This can be the case for fine dining operations with extensive menus and wine lists. In other operations, guest service training may consist primarily of providing employees with instructions for guest order taking and delivery using a drive-thru window. Every foodservice operation is unique, and so are its server training needs.

Regardless of the level of skills training required, however, foodservice operators have access to a wide variety are free-to-use server training programs. A large number of different entities now provide service-related training videos that can be helpful, and these videos can be easily accessed.

To examine the various types of server training videos that are free to use, go to YouTube. When you arrive at their site, enter "restaurant server training" in the search bar and review the results.

Generating Optimal Levels of Operating Profits

Whether it is a modest food truck, sub shop, or a fine-dining restaurant, every foodservice operation must manage to optimize its operating profits, and every foodservice operator must know and understand the components in the profit formula. That formula is:

Revenue – Expenses = Profit

When sales are made to guests, a foodservice operation receives revenue, and it also incurs the expenses required to generate the sales. Profit is the amount of money (if any!) that remains after all expenses have been paid. Since it is common in the foodservice industry, in this book the following terms will be used interchangeably: revenues and sales, expenses and costs, and profit and net income.

Key Term

Revenue: The term indicating the dollars taken in by a business within a defined time period. Also referred to as "sales."

Key Term

Expenses: The price paid to obtain the items required to operate a business. Also referred to as "costs."

Key Term

Profit: The dollars that remain after all a business's expenses have been paid. Also referred to as "Net income."

The profit formula holds true even for operators in the **nonprofit sector** of foodservice, and the key information presented in this book applies to all of them as well.

There are a number of names commonly used to identify nonprofit operations. These include noncommercial, institutional, contract feeding, on-site foodservices, and managed services. The number of nonprofit foodservice operations is vast and varied and includes food operations in:

Key Term

Nonprofit sector (foodservice): Foodservice in an organization where generating food and beverage profits is not the organization's primary purpose. Also referred to as the "noncommercial" sector.

K–12 education
Colleges
Universities
Hospitals
Nursing homes
Retirement centers
Correctional facilities
Military facilities
Businesses
Factories
Private clubs
Sporting arenas and stadiums
National and State parks
Transportation providers including airlines, trains, and cruise ships

The success of a commercial (for-profit) foodservice operation is most often determined by its sales volume (in dollars), and the resulting profits it generates. The success of nonprofit operations is often evaluated by participation rate (the number of people it serves). In all cases, however, revenue and expenses are just as important to nonprofit operators as they are to commercial operators. As a result, both commercial and nonprofit foodservice operators must attempt to optimize the difference between their revenue and their operating expenses if they are to be successful.

Find Out More

Many of the goals foodservice operators have for themselves and their businesses involve the effective control of operating costs and optimizing operating profits.

A close examination of the profit formula reveals that operators must pay attention to their revenue levels. Effective marketing, quality food and beverage products, and consistently outstanding service will all help to increase the size of this number. However, foodservice operators must also pay very close attention to their expenses (costs). Unless costs are properly controlled, profit levels will not be optimized. As a result, all foodservice operators must become "experts" at cost control.

The effective control of costs in a foodservice operation requires extensive knowledge and skill in a variety of areas. While there are many publications addressing "Cost Control" in foodservice operations, one of the best and most up-to-date cost control resources available to foodservice operators is the book, *Cost Control in Foodservice Operations*, part of the Wiley *"Foodservice Operations: The Essentials"* series.

To learn more about the content and availability of this extremely valuable John Wiley-published book, enter "Wiley Cost Control in Foodservice Operations" in your favorite search engine and review the results.

What Would You Do? 1.1

"I can't get her attention," said Roberto. "I keep trying, but she isn't looking over at us."

Roberto and his co-worker Adrianna were having lunch at Oscar's. Oscar's was a restaurant popular for its soups and sandwiches served in a casual setting.

"I know you asked for a side of mayonnaise when you ordered your sandwich. She must've forgotten," said Adrianna.

"I think you're right," said Roberto. "I didn't notice it when she delivered our food. Now I'm not sure exactly what to do. She hasn't checked back with us, and I don't see her."

"And your sandwich is getting cold," observed Adrianna.

Assume you were the operator of the Oscar's. What do you think will be Roberto's assessment of the sandwich quality he received during his visit? How does service quality directly impact a customer's view of product quality?

Foodservice Operators' Essential Tasks

To be successful, foodservice operators must effectively manage their businesses. What is **management**? A simple definition suggests that *"management is the process of using available resources to attain desired goals."*

A special concern about an organization's resources is that they are almost always limited, and few, if any, foodservice operators have all of their desired resources as they prioritize and pursue their goals.

Key Term

Management: The process of planning, organizing, directing, controlling, and evaluating the financial, physical, and human resources of an organization to reach its goals.

Examples of resources in a foodservice operation include money, facilities, equipment, employees, available time, and the work procedures that are in place. Other resources may include energy, inventories, and specialized items such as recipes and even specialized cooking techniques. Since resources are limited, however, foodservice operators must make good decisions about how they will be used.

There are numerous ways to view the principles of effective management. One good way is to consider the five primary functions that are essential to effectively managing any organization. These are shown in Figure 1.1.

Planning

This initial step in the management process addresses the creation of an operator's goals and objectives. This is the first step in the management process because it identifies precisely what a foodservice organization wants to achieve. It is important to use current and pertinent information and, when possible, participative input from staff members and guests who are affected by the plans. Plans must be flexible, and they must be implemented, evaluated, and changed when it is appropriate to do so.

Organizing

After objectives have been identified, an organization must ensure it has the funding, staff, equipment, and raw materials needed to achieve its objectives. These business **assets** must then be arranged (organized) in a way that optimizes the organization's ability to achieve its objectives.

Key Term

Asset (business): Property that is used in the operation of a business including money, real estate, buildings, inventories, and equipment.

Figure 1.1 The Management Process

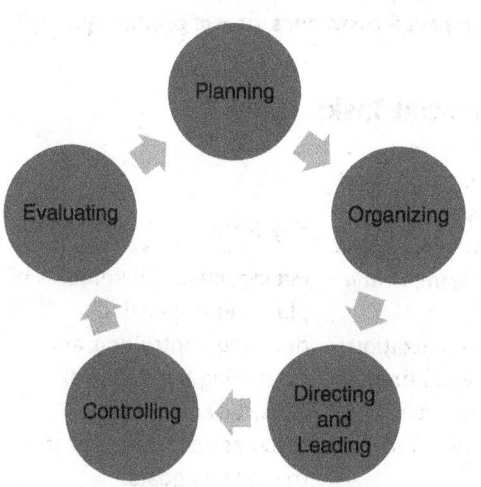

Directing and Leading

This important management function addresses the task of telling and showing all staff members exactly what is expected of them. When given clear directions, all staff members will know the important roles they will play in helping the organization achieve its objectives.

The best foodservice operators strive to be good leaders. While this is an easy goal to state, it is often much more difficult to achieve. What makes a good leader? While entire books (and bookshelves!) address this key question, a short list of effective leadership traits for foodservice operators includes:

✓ Using a good understanding of the operation's values and being able to translate these values into effective practices.
✓ Having an objective and measurable "picture" of the desired future of the operation.
✓ Helping others to develop the knowledge and skills needed to attain the operation's vision. This is done, in part, through effective employee selection, orientation, training and follow-up evaluation, and coaching activities.
✓ Developing a team of staff members who are committed to the operation's success.
✓ Achieving a reputation for quality that enables the foodservice operation to increasingly meet and exceed their guests' expectations.

Controlling

By continually assessing the work processes and procedures they have implemented, foodservice operators can better identify situations that might prevent their organizations from meeting objectives. In the foodservice industry, important processes that must be controlled include those activities related to purchasing, receiving, storing, preparing, and serving menu items, and later chapters in this book address each of these key tasks.

Evaluating

This final management activity requires that foodservice operators carefully assess their current performance and compare it to what has been planned or forecasted. This analysis helps to indicate where, if at all, corrective actions might reduce differences between planned for and actual financial results.

There are several steps involved in the evaluation process, and they can be summarized as:

1) Forecasting expected revenue and costs
2) Measuring actual operating results
3) Comparing expected and actual operating costs
4) Taking corrective actions as needed
5) Evaluating the effectiveness of the corrective actions

Each of these management functions requires foodservice operators to know and apply principles of effective management. As appropriate, these principles and the techniques required to put them into practice are addressed in the remaining chapters of this book.

Serving Food Safely

Serving high-quality menu items, providing service excellence, and optimizing profit levels are essential to the successful management of a foodservice operation. However, it is also essential to ensure the safety of the food items being served. These efforts affect those who serve them and those being served. This is not only the right thing to do, but it is also the legal thing to do.

All operations selling food to the public are subject to the requirements of the U.S. Food and Drug Administration (FDA) **food code**.

States and municipalities are responsible for ensuring that foodservice operations follow their own jurisdictions' versions of the food code. The FDA's food code assists the food control jurisdictions at all levels of government. One way to do this is to provide these governmental agencies with sound technical and legal bases for regulating the retail and foodservice segment of various industries. Examples include restaurants and grocery stores and institutions such as nursing homes and hospitals. Local, state, tribal, and federal regulators use the *FDA Food Code* as a model to develop or update their own food safety rules and to be consistent with national food regulatory policy. One primary objective of the food code is to assist foodservice operators in avoiding cases of **foodborne illness**.

One key requirement in the FDA food code, of which every foodservice operator must be aware, is the designation of a **person-in-charge (PIC)**.

The PIC is the designated individual present at a food establishment or operation who has the

Key Term

Food Code (FDA): A model for best practices to ensure the safe handling of food in retail settings.

Key Term

Foodborne illness: Foodborne illness is caused by consuming contaminated food, beverages, or water infected with a variety of bacteria, parasites, viruses, and/or toxins and poisons. These pathogens can be acquired through person-to-person spread, animal contact, the environment, and recreational or drinking water.

Key Term

Person-in-charge (PIC): The owner of the business, or a designated person—such as a chef, kitchen manager, or employee who is present at the work site and has direct authority and supervision over employees who engage in the safe storage, preparation, display, and service of food.

overall responsibility for the operation at the time. They oversee food preparation, supplies, quality of service, food safety, regulations, and techniques to maintain a food-safe environment.

During mandated food safety (health) inspections, most states require at least one PIC to be on the premises during operating hours. This PIC must be able to demonstrate knowledge of foodborne illness prevention and other factors at the request of a **health inspector**.

While each state is free to determine how its food safety inspections take place, foodservice operators are typically subject to three different inspection types. These are:

1) Routine inspections
2) Re-inspections
3) Complaint inspections

Key Term

Health inspector:
A professional responsible for ensuring compliance of sanitation codes required of businesses that are open to the public. This compliance is achieved through facility inspections that may be unannounced or occur as a result of a complaint.

Routine inspections are those periodic inspections that are performed as part of a jurisdiction's ongoing food safety system. Re-inspections are undertaken when a foodservice operation has committed violations of the food code that need correcting. Complaint inspections are performed in response to a complaint received by a member of the community.

Routine and complaint inspections are unannounced. This means that a foodservice operator is given no prior notice or prearranged time for the inspection to occur. Typically, if a re-inspection is required, the foodservice operation is given a specific date by which violations must be corrected, and this suggests a prearranged time for the revisit.

Depending upon the jurisdiction, after a health inspector has completed an inspection, the foodservice operation is given a score of satisfactory, unsatisfactory, or incomplete. A satisfactory score means there were no observed violations during the inspection or, if violations were observed, they were not significant enough to require correction before the next routine inspection. An unsatisfactory score means that the violations were significant enough to require correction before the next routine inspection. An incomplete inspection means that the inspector was interrupted in his or her work and had to leave before completing the inspection. In some jurisdictions, inspection scores (e.g. A, B, or C and the like, or a numerical "grade" of 1–100 score, or both) are given.

Two very important things foodservice operators must recognize about health inspection scores are that their PIC will accompany the inspector during the inspection and their scores are available to the public.

During an inspection, a PIC must demonstrate the required level of knowledge to be in compliance with the food code. If, during a routine inspection, the inspector finds no critical violations of the food code, the requirement for manager knowledge is met. A second way to demonstrate sufficient food safety knowledge is to be certified by an approved certification entity. To become certified, the PIC must have taken an accredited manager certification course that includes passing a test.

The third way to demonstrate manager knowledge is by correctly responding to an inspector's questions about the operation. During a routine inspection, the safe food practices and procedures used by the establishment will be discussed by the PIC and the inspector. The PIC must be able to show a working knowledge of those areas of the food code that directly relate to the foodservice operation. In all cases, the PIC must be qualified, and it is the foodservice operator's responsibility to ensure that is the case. Failure to do so will automatically result in an unsatisfactory inspection score.

The second and an increasingly important factor of which foodservice operators must be aware is that their inspection scores become part of the **public record**. That means that, in most jurisdictions, the inspection scores must be posted on-site for public view, or a copy of the most recent inspection report(s) must be available for public review when requested.

It is important that foodservice operators recognize that, in the past, a poor health inspection report was rather easily manageable. For example, negative exposure resulting from a poor inspection score did not extend beyond lost sales for the day and perhaps reputational issues for those who observe or are sickened by a food safety violation. Today the situation for foodservice operators is very different.

Key Term

Public record: Documents or other information that are not considered confidential, that generally pertain to the conduct of a government entity, and that can be viewed by the general population.

In most cases, it will only be several hours before a poor inspection report is published on a government's website. This is immediately followed by its distribution on websites and blogs that automatically collect such information about restaurant inspections. Some guests even use their smartphones to photograph posted inspection scores and then post those scores on their own social media sites.

Sometimes local media including newspapers and televised news produce segments aimed at sharing health inspection scores. Generally, the shared scores are poor, and the media entity will likely use dramatic tactics to present information about the foodservice operation. Negative publicity such as this can be significantly detrimental to the successful operation of a foodservice facility, and it must be avoided. Ensuring that an operation's menu items are always prepared and served in a safe manner is imperative when the operation is operated successfully.

Find Out More

Every foodservice operator must ensure they serve food safely. This requires knowledge in many areas including all aspects of foodservice that can affect food safety including:

1) Foodborne microorganisms & allergens
2) Personal hygiene
3) Purchasing, receiving, and storage
4) Preparation, cooking, and serving
5) Facilities cleaning/sanitizing and pest management
6) Regulatory authorities and inspection

The National Restaurant Association's ServSafe® Food Safety training program is the most widely recognized and respected in the foodservice industry because it:

✓ Uses quality materials and exams created by foodservice and regulatory experts exclusively for the foodservice industry
✓ Reinvests proceeds from programs back into the industry
✓ Is accepted in all 50 states, making it ideal for single and multi-unit operations
✓ Is a single source, one-stop show for both food safety training and the certification examination
✓ Delivers up-to-date regulatory information
✓ Provides support from foodservice subject matter experts available to answer questions
✓ Offers flexible online, classroom, in-unit and one-on-one training and examination options

To learn more about this important operating aid and food safety training tool, enter "NRA ServSafe" in your favorite search engine and review the results.

Serving Nutritious Food

A foodservice operator need not be a **registered dietitian (RD)** to recognize that it is important to serve nutritious food.

Increasingly, foodservice guests make decisions about which operations to frequent based on personal nutritional and dietary concerns. As previously stated, all of a foodservice operator's

Key Term

Registered Dietitian (RD):
A health professional who has special training in diet and nutrition. RDs offer advice on nutrition and healthy eating habits to help people improve their health and well-being.

menu items must be made with wholesome food that is safe to consume, and the items served must be nutritious as well. Doing so requires a conscientious effort on the part of foodservice operators and their entire staff.

Nutritious food is prepared in a way that optimizes retention of the food's **nutrients**. The serving of nutritious food does not happen by accident. Rather, nutrition concerns must be addressed in a number of key areas including:

Key Term

Nutrient: A chemical compound such as protein, fat, carbohydrate, vitamin, and/or minerals contained in foods that are used by the body to help it function and grow.

✓ Menu planning
✓ Menu design
✓ Purchasing
✓ Receiving and storage
✓ Menu item production and service

Experienced foodservice operators recognize that their guests are increasingly concerned about the nutritional value of the menu items they purchase. Many guests make their dining out decisions based on their nutrition-related perceptions of the operations they want to visit.

Some guests who have never visited or even heard about a foodservice operation may learn about its nutritious menu item alternatives and may want to visit the property. Increasingly prospective guests utilize the Internet to search for new dining out options. Increasingly as well, familiar names in the foodservice industry are placing a substantial amount of nutrition-related information on their websites. Some of this information is required by law for certain operators, but much of it is not.

For example, there is no requirement that foodservice operators indicate which of the food items they sell are made from **organic** food ingredients. Yet, for a variety of reasons, including the support of local agriculture, avoiding foods with pesticides, and supporting those producers who create products in ways that are good for the environment, an ever increasing number of guests will be interested in whether a foodservice operation serves organic products and will prefer to frequent those businesses who do.

Other examples of guests seeking products based on their perceptions of nutrient value include the increasing interest in plant-based

Key Term

Organic (food): the product of a farming system which avoids the use of synthetic fertilizers, pesticides, growth regulators, and livestock feed additives. The U.S. Department of Agriculture (USDA) requires that food products sold, labeled, or represented as organic in the United States must have at least 95% certified organic content.

milks (e.g. oat, almond, soy, and coconut milks) and the increased availability of plant-based meat substitutes.

Foodservice operators concerned with serving nutritious foods must direct their attention both to the products they buy and how these products are handled and prepared after they have been received. For example, a foodservice operation may advertise that it serves only organic vegetables. However, boiling and cooking some vegetables in high temperatures or in excessive amounts of water decrease their nutrient level. Water soluble vitamins like Vitamins C and B are often lost during these types of cooking methods. Minerals including potassium, phosphorus, calcium, magnesium, iron, and zinc may also be significantly reduced when improper cooking methods are used.

Serving nutritious foods to guests can be especially challenging to foodservice operators because not all guests have the same perception about what does and does not constitute nutritious food. To cite an example, many bakers make their crusts for fruit pies using lard. Lard and butter are both used in baked goods, but some properties of lard make it a favorite for many bakers. For example, the melting point of lard is lower than butter, so more air and steam are released during baking times. The result: greater leavening and a flakier texture in baked goods prepared with lard.

If a server is asked by a guest whether an operation's cherry pie crust was made with lard and the server responds with an enthusiastic "Yes!" the guest's reaction may be favorable because this guest knows a superior baked product will be prepared.

Other guests, however, may know lard is a hydrogenated fat (also called a trans-fat), and that consumption of trans-fats is linked to an increased possibility of heart diseases. These guests may decide *not* to buy the pie because it's crust was made with lard. In this case, a well-trained server may be able to suggest dessert alternatives to those guests.

Regardless of the ingredients used to make their foods, and the preparation methods utilized, all foodservice operators must recognize that **accuracy in menu laws** require that descriptions of the menu items they sell are truthful and accurate (see Chapter 6).

Nutrition Essentials for Back-of-House Staff

With some exceptions, back-of-house staff are those primarily involved with production of the menu items sold in a foodservice operation.

Key Term

Accuracy in menu laws: Legislation that requires foodservice operations to truthfully, and accurately, represent the quality, quantity, nutritional value, and price of the menu items they sell.

Also known as "Truth-in-menu laws" and "Truth-in-dining laws."

Special nutritional concerns about which back-of-house staff must be aware (knowledgeable) include:

1) *Understanding the menu:* Production staff must be aware of menu items that meet the needs of guests with specialized nutritional concerns. Examples include knowing which menu items are gluten-free, low-fat, or low-sodium.

2) *Maintaining nutritional quality:* Production staff must be knowledgeable about the purchasing, storing, and production techniques required to maximize the nutritional content of the menu item being produced.

Key Term

3) *Following standardized recipes:* A **standardized recipe** details the procedures used to prepare and serve each of an operation's food or beverage items. A standardized recipe ensures that each time guests order a menu item it has been prepared the same way with no preparation "short-cuts."

Standardized recipe: The instructions needed to consistently prepare a specified quantity of food or drink at an expected quality level.

From a nutritional perspective, the information provided about a menu item will only be accurate if the recipe used to produce the item is carefully and consistently followed. Substitution of ingredients or changes in quantities of ingredients used to make an item, as well as variations in portion size will affect the nutritional content of that item. For that reason, it is essential that food production staff always carefully follow standardized recipes.

4) *Recognizing characteristics of quality ingredients:* Many of a foodservice operator's menu items are produced using perishable products that deteriorate during storage. As a result, it is especially important that back-of-house employees be trained adequately to recognize quality products and substandard products, so that the latter are not utilized when making menu items.

5) *Understanding the importance of serving size:* There is no requirement that a foodservice operation must serve a specific-sized portion of a menu item. However, nutritional information provided about a menu item will be based, in part, on the size of portion to be served. For that reason, it is important that food production personnel serve only portions of management-approved size and specified by the standardized recipe.

Nutrition Essentials for Front-of-House Staff

An operation's front-of-house staff may have a limited role in producing some menu items, but primarily they are responsible for the proper service of the menu items sold. Service staff require nutritional knowledge and skills so they can be integrated into proper guest service. Special nutritional concerns about which front-of-house staff must be aware (knowledgeable) include

1) *Understanding the menu:* As is true with production personnel, service staff should learn basic ingredients and preparation methods for all the items they serve. Service staff should be able to provide basic nutrition-related information to guests who request it. For example, a fish dish listed on a menu may tell what type of fish it is, but health-conscious guests may want to know if the item is pan-fried, broiled, or sauteed. Servers must be able to provide this information.

   To cite a second example, many guests are concerned about gluten in foods. Some are gluten intolerant or may have Celiac disease (an immune reaction to eating gluten which is a protein found in wheat, barley, and rye.) For these guests, it is important that servers know gluten-free foods include:

   - Most dairy products such as cheese, butter, and milk
   - Fruits and vegetables
   - Meat and fish (although not breaded or battered)
   - Potatoes
   - Rice and rice noodles
   - Gluten-free flours including rice, corn, soy, and potato flour

   The nutrition-related knowledge required of foodservice servers will vary based on the operation in which they are employed. However, in all cases staff members serving the public will often be called upon to answer basic nutrition-related questions, and they must be prepared to do so.

2) *Making appropriate substitutions:* In some foodservice operations, guests may be able to modify standard menu items with reasonable requests that must be communicated by a server to the operation's production staff. For example, a breakfast guest who has ordered fried eggs, potatoes, and bacon may inquire about substituting fruit for the toast that normally comes with their breakfast selection. In situations such as this, servers must be instructed to respond to special substitution requests so guest service levels can be optimized.

3) *Making appropriate recommendations:* Proper menu design and menu item descriptions (see Chapter 6) can be extremely helpful in informing guests about some of the nutritional aspects of menu items sold. However, servers may be asked for recommendations about menu items that may be low-calorie, low-fat, or low sodium. A properly trained server should be able to make recommendations that accommodate their guests' preferences.

Technology at Work

Since 2018, the Food and Drug Administration (FDA) has mandated that consumers have easy access to calorie and nutrition information in certain chain establishments covered by the rule. For example, the menu labeling requirements apply to restaurants and similar retail food establishments that are part of a chain with 20 or more locations.

Covered establishments must disclose the number of calories contained in standard items on menus and menu boards. Businesses must also provide (upon request) the following written nutrition information for standard menu items: total calories; total fat; saturated fat; trans-fat; cholesterol; sodium; total carbohydrates; sugars; fiber; and protein. In addition, two statements must be displayed. Operators must indicate that this written information is available upon request. The second statement relates to daily calorie intake and must indicate that 2,000 calories a day is used for general nutrition advice, but calorie needs vary between individuals.

Even smaller foodservice operations that are not required to follow this mandate have found that their guests now ask them for information similar to that provided by chain operators. Guests increasingly demand more transparency in what they eat, how it is prepared, and the nutritional profile of the items they buy. In response, foodservice operators must adapt and showcase the healthy aspects of the items they sell.

Providing nutrition-related information can be challenging for small operators. However, nutrition label software (also known as nutritional analysis software) allows even small restaurants to provide their guests with key information about their menu items. A nutrition label software program can provide a complete breakdown of calorie content, fat, protein, and carbohydrates, among other information.

Nutrition label data is increasingly important for consumers on special diets, those with food allergies, or those looking to improve their health using food nutrition information. To learn more about the features and costs of utilizing these nutrition-information programs enter "nutrition label software for restaurants" in your favorite search engine and view the results.

This chapter addressed several key areas requirements for success in a foodservice operation. There can be different ways to measure the success of a foodservice operation for both commercial and noncommercial operators. However, recording and reporting data about their income and expenses are always a key component of success assessment.

Properly recording and reporting the financial results of a foodservice operation in a way that can be easily understood by others is such an important skill for foodservice operators. They want to be successful, and accounting and financial management principles needed to accomplish this are the topic of the next chapter.

What Would You Do? 1.2

"That's the third customer today that left without buying anything," said Isabella.

Isabella was talking to Mateo, the owner of The Tasty Bakery; a small bakeshop that specialized in premium-quality cakes, cookies, muffins, and pastries.

"Did they think our prices were too high?" asked Mateo.

"No, they didn't seem to have any problem with the prices," replied Isabella, "They wanted a dozen gluten-free chocolate chip cookies for an office party this afternoon. I told her the only gluten-free item we sold was our blueberry muffins. The other two customers who left without buying anything today were looking for gluten-free cookies as well. They didn't seem mad, they just left without buying anything."

Assume you were Mateo. How important do you think it would be for you to carefully monitor your customers' requests for specialized items they want to buy from your shop but cannot purchase them? Do you think the customers who left without buying anything were more concerned about nutritional issues or the prices you charge for the items you sell?

Key Terms

Quick service
    restaurants (QSRs)
Value
Service industry
Front-of-house staff
Back-of-house staff
Revenue
Expenses

Profit
Nonprofit sector
    (foodservice)
Management
Asset (business)
Food Code (FDA)
Foodborne illness
Person-in-charge (PIC)

Health inspector
Public record
Registered Dietitian (RD)
Nutrient
Organic (food)
Accuracy in menu laws
Standardized recipe

Operator's 10-Point Tactics for Success Checklist

Evaluate your need for, and the current status of, each of the following operational tactics. For those tactics you think are important, but not yet in place, develop an action plan for its implementation including who will be responsible for the tactic's completion and the target date by which it should be completed.

Tactic	Don't Agree (Not Done)	Agree (Done)	Agree (Not Done)	If Not Done	
				Who Is Responsible?	Target Completion Date
1) Operator recognizes that they are one of the few types of professional managers who are responsible both for producing a product and delivering it to its end-user (guests).	____	____	____		
2) Operator understands the importance to success of serving high-quality products.	____	____	____		
3) Operator understands the importance to success of providing guests with high-quality service.	____	____	____		
4) Operator understands the importance to success of generating optimal levels of operating profits.	____	____	____		
5) Operator recognizes that planning, organizing, directing and leading, controlling, and evaluating are essential tasks in the operation of a successful foodservice business.	____	____	____		
6) Operator recognizes the important role played by the person-in-charge (PIC) in the successful operation of a foodservice business.	____	____	____		

				If Not Done	
Tactic	Don't Agree (Not Done)	Agree (Done)	Agree (Not Done)	Who Is Responsible?	Target Completion Date
7) Operator can state the differences between the three types of health inspections to which their operations are subject.	____	____	____		
8) Operator understands the importance to success of serving nutritious food.	____	____	____		
9) Operator can identify key nutrition issues that relate to back-of-house staff.	____	____	____		
10) Operator can identify key nutrition issues that relate to front-of-house staff.	____	____	____		

2

Accounting and Financial Management in Foodservice Operations

What You Will Learn

1) The Importance of Accounting and Financial Management
2) How to Prepare and Read an Income Statement
3) The Goal of Break-even Point Analysis

Operator's Brief

In this chapter, you will learn that professional accounting methods must be used to record and report the financial results of operating your business. Managerial accounting is the accounting specialization that uses historical information to estimate future operating results so a foodservice operation can manage its money and achieve its financial goals.

Professional accounting in a foodservice operation consists of three main tasks:

1) Bookkeeping
2) Summary accounting
3) Financial analysis

In this chapter, you will learn how each of these tasks can help you prepare accurate and timely financial summaries of your operating results.

While there is no requirement that it be used, many foodservice operators utilize the Uniform System of Accounts for Restaurants (USAR) to prepare their financial documents and, in this chapter, you will learn the advantages of doing so.

Your income statement is one of your most important financial summaries. It details, for a specific accounting period, an operation's revenue from all revenue-producing sources. It also reports the expenses required to generate the revenues and any resulting profits or losses. In this chapter, you will learn how an income statement is prepared using the USAR format.

A USAR-formatted income statement provides information about seven key areas in a foodservice operation. These are:

1) Sales (Revenue)
2) Total Cost of Sales
3) Total Labor
4) Prime Cost
5) Other Controllable Expenses
6) Non-controllable Expenses
7) Income (Profits)

In this chapter, you will learn why understanding each of these seven areas is important. You will also learn about EBITDA, a measure that business owners and operators use to determine their net cash earnings (before taxes) during a specific accounting period.

Finally, in this chapter you will learn about the importance of break-even point analysis: the point in a foodservice operation when the level of sales generated is exactly equal to expenses incurred. Understanding the break-even point is essential because, with sales above your break-even point, you will generate a profit, and with sales below your break-even point you will generate an operating loss.

CHAPTER OUTLINE

Professional Accounting
    Accounting for Successful Financial Management
    The Mechanics of Accounting
    Summary Accounting
    Financial Analysis
    The Uniform System of Accounts for Restaurants (USAR)
    Sales (Revenue)
    Total Cost of Sales
    Total Labor
    Prime Cost
    Other Controllable Expenses
    Noncontrollable Expenses
    Profits (Income)

Professional Accounting

Some foodservice operators believe **accounting** for their businesses is an extremely complex process. While professional accounting does require attention to detail, every foodservice operator can master the basic skills required to effectively record, analyze, and manage their financial information successfully.

The term "accounting" originated from an old Middle French word, "*acompter*," which originated from Latin ("*ad +compter*") meaning "to count." Since all foodservice operators know how to count, they can all master the most important accounting principles even if they do not consider themselves to be a professional **accountant**!

Key Term

Accounting: The system of recording and summarizing financial transactions and analyzing, verifying, and reporting the results.

Key Term

Accountant: An individual skilled in the recording and reporting of financial transactions.

Accounting for Successful Financial Management

The purpose of professional accounting is to report (account for) an operation's money and other valuable property. In a foodservice operation, accounting principles are utilized every time a guest purchases food or beverages. Accounting and financial management in a foodservice business, however, occurs even before the operation opens. Consider that operators estimate their initial costs before they decide to open their business, and they often seek loans from banks or others to provide them with funds needed to begin their business.

Accounting-related information is important in all phases of a foodservice business's planning, development, and operation. The goal of accounting is to provide operators with useful information. To be useful, the information must be accurate and timely. To be most valuable, it must also be easy to understand and use.

To understand why accounting plays such a significant role in business, consider just a few examples of the type of basic and important questions the use of accounting information can answer for foodservice operators:

1) What was the total sales level achieved by our business last month?
2) How many guests did we serve?
3) What was our most popular menu item?
4) What percentage of our revenue was achieved from sales of alcoholic beverages?

5) What was the **guest check average** achieved last week? Was it higher or lower than the prior week?

6) What portion of our revenue is being spent on labor?

7) What percentage of our sales was achieved through our take-out and delivery services?

8) Are our operating expenses higher or lower than those of other operations of a similar type?

Key Term

Guest check average: The average (mean) amount of money spent per guest (or table) during a specific time. Also referred to as "check average" or "ticket average."

9) Are we more (or less) profitable this month than last month?

10) What is our operation realistically worth if we were to sell it today?

There are a variety of different accounting specialty areas including tax accounting, cost accounting, and financial accounting. **Managerial accounting** is the specialty area of accounting that is the primary topic of this book. It is the specialization that helps managers make decisions about the future. To better understand the purpose of managerial accounting, assume that Rebeka Gautier is responsible for providing in-flight meals to international travelers on flights from New York City to Paris. She manages a large commercial kitchen located near John F. Kennedy Airport.

Key Term

Managerial accounting: The accounting specialty that uses historical and estimated financial information to help foodservice operators plan the future.

Rebeka's clients are the airlines who count on her operation to provide passengers with tasty and nutritious meals at a per-meal price airlines find affordable.

One of Rebeka clients wishes to add a new daily flight beginning next month. The evening flights will carry an average of 500 travelers, each of whom will be offered one of two in-flight meal choices for dinner. The client would like to provide each flier with a choice of a beef or a chicken entrée. To ensure that the maximum number of fliers can receive their first choice, should Rebeka's operation provide each flight with 500 beef and 500 chicken entrées? (The answer, most certainly, is "No!")

To prepare 1,000 meals (500 of each type) would ensure that each traveler always received his or her first meal choice. However, it would also result in the production of 500 wasted meals (the 500 meals *not* selected) on each flight. It would be difficult for Rebeka to provide the airline with cost-effective per-meal pricing when this number of meals is inevitably wasted.

The more cost-effective approach would be to accurately forecast the number of beef and chicken entrées that would likely be selected by each group of passengers and then produce that number. The problem, of course, is in knowing the optimum number of each meal type that should be produced. If Rebeka had carefully and consistently recorded previous meal-related transactions (entrées chosen by fliers on previous flights), she would be in a much better position to use this information. She could, for example, estimate the actual number of each entrée the new passengers will likely select. If she had done so, she would be using managerial accounting.

Managerial accounting is the system of recording and analyzing transactions for the purpose of making management decisions of this kind. It consists of utilizing accounting information (historical records in this case) to make informed management decisions.

Managerial accounting is one of the most exciting accounting specializations. Its proper use requires skill, insight, experience, and intuition, and these are the same characteristics possessed by the best foodservice operators. As a result, successful foodservice operators most often become excellent managerial accountants.

The Mechanics of Accounting

As shown in Figure 2.1, the financial management of a foodservice operation consists of three key components. It is essential that all foodservice operators understand the purpose and importance of each component.

Bookkeeping

A foodservice operation's accounting management system must begin with proper **bookkeeping**. Successful bookkeeping requires the accurate recording of every business transaction that occurs in a foodservice operation. Examples of

Key Term

Bookkeeping: The process of recording a foodservice operation's financial transactions into organized accounts each day.

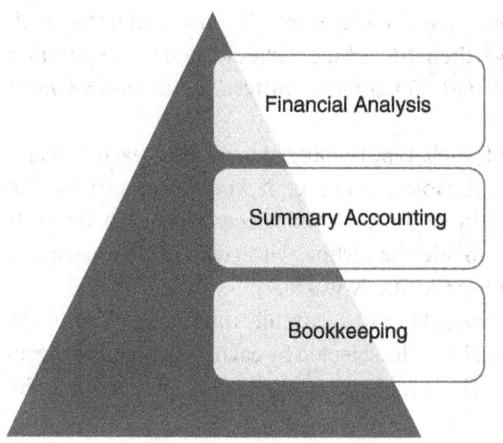

Figure 2.1 The Mechanics of Financial Management

Financial Analysis

Summary Accounting

Bookkeeping

transactions that must be recorded include income received from the sale of menu items to guests and payments made to an operation's employees, vendors, and suppliers.

Proper bookkeeping provides the foundation of accurate financial reporting and analysis. It is not possible to make a meaningful analysis of a foodservice operation's financial standing if the data to be summarized was erroneously or carelessly supplied by those performing bookkeeping tasks. In addition, foodservice managers will not be able to properly analyze financial information and make correct decisions if the accounting information summarized was inaccurate.

To better understand the financial management process, consider that recording an individual financial transaction such as the sale of a cup of coffee is a bookkeeping task. It is typically completed by a server using an operation's **point-of-sale (POS) system**.

An operator may then record the total number of cups of coffee sold in a specified **accounting period** based on POS system-supplied information. These sales and others achieved by the operation will be summarized in financial statements to be analyzed by the operation's accountants, operators, and owners.

In the foodservice industry, the actual distinctions between bookkeeping and summary accounting are not always clear. Many foodservice operations are small, and bookkeeping, summary accounting, and financial analysis (managerial accounting) may all be done by only one or two

Key Term

Point-of-sale (POS) system: An electronic system that records foodservice customer purchases and payments and other operational data.

Key Term

Accounting period: The amount of time included in a financial summary or report that should be clearly identified on the financial document. For example, a week, month, or year.

individuals. For this reason, this book will not make a significant distinction between bookkeeping and accounting. For foodservice operators, however, it is important to ensure that precise and timely bookkeeping (recording) methods are used to produce the accurate financial data needed to make good managerial decisions.

Summary Accounting

Foodservice operators should regularly create financial summaries of the transactions that have occurred in their businesses. These specific summaries vary based on the needs and desires of the business's owner. The three most useful and commonly prepared summary financial statements for a foodservice operation are:

✓ Income statement
✓ Balance sheet
✓ Statement of cash flows (SCF)

The **income statement** is the financial summary used by foodservice operators to document and report their income and expenses over a specific accounting period. For the time addressed in the summary, the document tells foodservice operators how much revenue was generated, how much expense was incurred to generate the revenue, and the dollar amount of profit (if any!) that was achieved.

The purpose of the **balance sheet** is to report the financial status of a business at a specific point in time. The balance sheet shows what a business owns (assets), how much it owes (liabilities), and the amount its owners have invested in the business (equity). The basic accounting equation used to create a balance sheet is:

$$Assets = Liabilities + Owner s \ Equity$$

The balance sheet is most useful when it is prepared and compared over several consecutive periods so trends in its different components can be analyzed.

The purpose of a **statement of cash flows (SCF)** is to report what has happened to a business's cash balances during a specified accounting period. This financial summary helps to demonstrate a business's ability to operate in the short- and long-term based on how much cash is flowing in and out of the business.

Foodservice operations can generate (use) cash from three specific types of activities:

1) Operating
2) Investing
3) Financing

Operating activities detail cash flow that is generated from a business's normal operations. Investing activities include cash changes resulting from buying or selling assets such as real estate or large equipment. Financing activities detail cash flow from acquiring debt and repaying borrowed money.

Key Term

Income statement: Formally known as "The Statement of Income and Expense," a report summarizing a foodservice operation's profitability including details regarding revenue, expenses, and profit (or loss) incurred during a specific accounting period. Also commonly called the Profit and Loss (P&L) statement.

Key Term

Balance sheet: A report that documents the assets, liabilities, and net worth (owner's equity) of a foodservice business at a single point in time. Also commonly called the Statement of Financial Position.

Key Term

Statement of Cash Flows (SCF): A report providing information about all cash inflows a company receives from its on-going operations and external investment sources. It also includes all cash outflows paying for business activities and investments during a defined accounting period.

It is important to note that cash flow is different from profit. That is why a SCF is most often interpreted together with other financial documents including the income statement and the balance sheet. Ideally, a foodservice operation should generate a **positive cash flow**. That is, the amount of cash on hand should be increasing, (not decreasing) as the foodservice business continues to operate.

Financial Analysis

The third and last tool illustrated by Figure 2.1 is the Financial Analysis. The regular and correct analysis of a foodservice operation's financial summaries are keys to a foodservice operation's success. Foodservice operators frequently use a five-step process to analyze the financial performance of their businesses.

Step 1: Performance standards (expectations) are established. This is typically done through the development of an operating budget (see Chapter 12).
Step 2: Actual financial information is collected and summarized to measure operating results.
Step 3: Comparisons are made between expected performance (Step 1) and actual performance (Step 2).
Step 4: Corrective action must be taken when necessary to identify causes and bring actual results (Step 2) in line with the expected performance (Step 1).
Step 5: An evaluation of the results of corrective actions that were taken (Step 4) is made.

Key Term

Positive cash flow: The condition that exists if the amount of a business's cash that is acquired exceeds the amount of cash spent by the business. Stated differently, positive cash flow means more cash is coming into a business than is going out of the business, and this is essential for the business to sustain long-term growth.

Technology at Work

There are numerous high-quality accounting software packages available for use by foodservice operators. One of the most popular is Intuit's "QuickBooks." Over 29 million businesses in the United States use QuickBooks to manage their accounting. In the restaurant industry, QuickBooks is the overwhelmingly preferred choice of accounting software, and it has roughly an 80% market share.

QuickBooks is a comprehensive accounting and business management program designed for small- and medium-sized businesses. Using QuickBooks,

(Continued)

foodservice operators can automatically generate financial summaries such as income statements, balance sheets, statements of cash flows (SCF), and more.

To learn more about the various product options and pricing packages available from QuickBooks, enter "QuickBooks for restaurants" in your favorite search engine and review the results.

* https://www.plateiq.com/blog/quickbooks-restaurant-accounting/ retrieved April 20, 2023

The Uniform System of Accounts for Restaurants (USAR)

Different businesses have different accounting needs, and there are systems of accounting produced specifically for individual business segments. In the hospitality industry, some of the best known of these uniform systems include the:

✓ **Uniform System of Accounts for Restaurants (USAR)**
✓ Uniform System of Financial Reporting for Clubs (USFRC)
✓ Uniform System of Accounts for the Lodging Industry (USALI)

Key Term

Uniform System of Accounts for Restaurants (USAR):
A recommended and standardized (uniform) set of accounting procedures used for categorizing and reporting restaurant revenue and expenses.

Uniform accounting systems are continually reviewed and periodically revised. For example, this book was prepared using reporting principles contained in the eighth and most current edition of the USAR. Important specific recommendations of the USAR will be addressed in detail in the appropriate portions of this book.

To illustrate why the use of the USAR is so important, assume that an individual owned two Italian restaurants located in two different cities, and both offered the same menu. The owner wants to assess the ability of the two persons (General Managers) that operate the restaurants. The reason: It would be very confusing if the units' two operators used different methods for preparing and reporting each of their operations' financial results. In this example, unless both operators report and account for their financial performance in a way that is consistent (uniform), the performance of the two operations and their operators could not be properly analyzed and/or compared to each other.

The use of the USAR to produce accounting information for foodservice operations is not mandatory, but use of the USAR is highly recommended. One reason is that the primary purpose of preparing accounting information is to clearly identify revenue, expenses, and profits for a specific time. The best foodservice operators want to do this properly so the financial records of their businesses will accurately reflect their efforts and success.

The USAR provides detailed instructions on how to produce an income statement, which is one of a foodservice operator's most important financial management documents.

The USAR Income Statement

All commercial foodservice businesses want to be profitable, and foodservice owners and their managers want to make a profit. As previously mentioned, the tool operators use to document and report their profits is the income statement which is formally known as "The Statement of Income and Expense." However, some operators refer to the income statement as the "profit and loss" (P&L) statement and, while that name is in common use, this book refers to the document by its shortened name: the income statement.

A foodservice operation's revenue minus its expenses equals its profits. The income statement details, for a specific accounting period, an operation's revenue from all revenue-producing sources. It also reports the expenses required to generate the revenues, and the resulting profits or losses.

The following is the basic profit formula (see Chapter 1), and it is also the format followed when preparing an operation's income statement.

Revenue – Expenses = Profit

To illustrate the use of the formula in preparing an income statement, consider Jerry, the operator of Jerry's Restaurant. His operation generates revenues from a variety of sources:

✓ In-house food sales
✓ In-house alcoholic beverage sales
✓ Gift card sales
✓ Carry-out sales
✓ On-site banquet (catering) sales
✓ Off-site banquet (catering) sales

In addition to generating revenue, Jerry creates expenses and incurs costs as he operates each of the major revenue centers. His goal should be to generate a profit in each of the revenue sources he manages. In fact, many foodservice operators consider each individual revenue-generating segment within their business to be a **profit center**, a term coined in 1945 by international management consultant Peter Drucker.[1]

Key Term

Profit center: A part of a business organization with assignable revenues and expenses. Also referred to as a "revenue center."

---

1 https://en.wikipedia.org/wiki/Profit_center#:~:text=Peter%20Drucker%20originally%20coined%20the,cheque%20hasn't%20bounced%E2%80%9D retrieved July 22, 2023

The *Revenue – Expenses = Profit* formula is also applicable to what is not typically considered a for-profit segment of the foodservice industry. For example, consider the operation managed by Harriet Bentevina.

Harriet is the manager of the employee cafeteria at Consolidated Industries, a large international corporation employing over 2,000 workers at its company headquarters. She provides employee meals at no cost to a large group of employees and managerial staff working at the company.

In this situation, Harriett's is not required to make a "profit." In fact, many noncommercial foodservice operations provide meals as a service to employees, students, patients, military, or others as a no-cost or greatly reduced-price benefit. Therefore, in these situations, operators manage a **cost center** that generates expenses, but with little or no direct revenue paid by the meals' consumers.

> **Key Term**
>
> **Cost center:** A part of a business organization with assignable expenses that generates little or no revenue.

In the noncommercial segment of the foodservice industry, an operator's goal may be to maintain or not to exceed a predetermined cost for operating the cost center. Whether they are operating a profit center or a cost center, however, all foodservice operators must know as much as possible about how their revenues and expenses are generated to best maximize their income and control their costs. The income statement is the primary tool foodservice operators use to gather and assess this necessary revenue and expense information.

The results reported on a foodservice operation's income statement are important to several different groups. The owners of a business typically have the greatest interest in its success because they reap the profits generated by the business. In many foodservice operations, however, the owners do not actively participate in managing the business. By evaluating income statement data, owners can better determine the effectiveness of the operators they have selected to manage their businesses, and the progress that has been made to achieve the owner's financial operating goals.

Lenders and potential investors will also be interested in the financial results shown on an income statement. Lenders will want to know if loans they make will be repaid in a timely manner, and potential investors will want to know if they will likely achieve their desired **return on investment (ROI)**.

> **Key Term**
>
> **Return on investment (ROI):** A measure of the ability of an investment to generate income.

Perhaps most importantly for themselves, foodservice operators consider the income statement to be the best reflection of their managerial ability. This is true for several reasons. First, it details how profitable an operation has been within a designated time. As well, the best managers usually operate more profitable facilities than do other managers. Operational performance as measured by the income statement's results is often used to establish managers' pay raises and compute their bonuses. As well, performance can be important when promotional opportunities are determined. Therefore, managers must be able to read and understand well their business's income statements.

An income statement can be prepared for any accounting period determined by a foodservice operator. However, the accounting periods should make good sense for the business to which they are applied. For many businesses, accounting periods coincide with the calendar months of the year (calendar year accounting period, or any consecutive 12-month period (fiscal year accounting period). Many (but not all) businesses produce a calendar year income statement because it eases the operation's ability to file annual tax returns, which are often based upon the calendar year.

In addition to preparing their annual income statement, most foodservice operators also prepare shorter accounting period income statements. Sometimes, they address a 4-month (quarterly) accounting period, but more often they are created monthly. In some cases, a foodservice operator may prefer to create income statements that are 28 days long because they want to create "equal" **28-day accounting periods**.

Key Term

28-day accounting period: An accounting period that is four weeks (28 days) in length instead of a calendar month that has between 28 and 31 days. There are 13 four-week periods instead of 12 monthly periods when using this system.

When preparing income statements based on a 28-day accounting period, each period is equal in length and will have the same number of Mondays, Tuesdays, Wednesdays, and so forth. This helps the operator compare performance from one accounting period to the next without having to compensate for "extra days" in any one period.

Figure 2.2 summarizes the most popular lengths of time used by foodservice operators and their accountants to create income statements.

Accounting Period	Number of Days Included
Month	Varies
Quarter	Varies
Annual	365 days (except leap year)
28-day	28 days
Weekly	7 days
Daily	1 day

Figure 2.2 Common Income Statement Accounting Periods

There are a variety of ways that foodservice operators *could* report their revenue, expense, and profits on an income statement. The USAR has recommendations for how an income statement *should* be produced. The USAR recommendations are not mandatory, but they are highly recommended and are followed by most foodservice operators and by all hospitality accounting firms.

To illustrate the production of an income statement using the USAR recommendations, consider Figure 2.3. It shows the USAR suggested income statement format used by Jerry's Restaurant for the year ending December 31, 20xx. Note: The "Line" column has been added by the authors for the reader's ease in identifying data locations.

Jerry's Restaurant

Income Statement

For the Year Ended December 31, 20xx

Line		
1	SALES	
2	Food	$ 1,891,011
3	Beverage	$ 415,099
4	**Total Sales**	**$2,306,110**
5	COST OF SALES	
6	Food	$ 712,587
7	Beverage	$ 94,550
8	**Total Cost of Sales**	**$ 807,137**
9	LABOR	
10	Management	$ 128,219
11	Staff	$ 512,880
12	Employee Benefits	$ 99,163
13	**Total Labor**	**$ 740,262**
14	PRIME COST	$ 1,547,399
15	OTHER CONTROLLABLE EXPENSES	
16	Direct Operating Expenses	$ 122,224
17	Music and Entertainment	$ 2,306
18	Marketing	$ 43,816
19	Utilities	$ 73,796

Figure 2.3 Sample USAR Income Statement

Line		
20	General and Administrative Expenses	$ 66,877
21	Repairs and Maintenance	$ 34,592
22	**Total Other Controllable Expenses**	**$ 343,611**
23	**CONTROLLABLE INCOME**	**$ 415,100**
24	**NONCONTROLLABLE EXPENSES**	
25	Occupancy Costs	$ 120,000
26	Equipment Leases	$ –
27	Depreciation and Amortization	$ 41,510
28	**Total Noncontrollable Expenses**	**$ 161,510**
29	**RESTAURANT OPERATING INCOME**	**$ 253,590**
30	Interest Expense	$ 86,750
31	**INCOME BEFORE INCOME TAXES**	**$ 166,840**

Figure 2.3 (Continued)

The USAR income statement is arranged on the above income statement *from the expenses that are most controllable to least controllable* by a foodservice operator. The format of a USAR income statement can best be understood by dividing it into its main sections:

1) Sales (Revenue)
2) Total Cost of Sales
3) Total Labor
4) Prime Cost
5) Other Controllable Expenses
6) Noncontrollable Expenses
7) Income (Profits)

Sales (Revenue)

After listing the name of an operation and the accounting period addressed, the USAR format for an income statement lists sales (Line 1), or revenue, first and on the top line. **Topline revenue** represents the total sales generated by a business before any expense deductions.

Key Term

Topline revenue: Sales or revenue shown on the top of the income statement of a business.

Historically, food sales (Line 2) include income from the sale of all food items and from the sale of nonalcoholic beverages such as soft drinks, coffee, tea, milk, bottled water, and fruit juices. When alcoholic beverages are sold, the revenue from the sales of these beverage products (Line 3) is added to food sales to yield an operation's **total sales** (Line 4).

Key Term

Total sales: The sum of food sales and alcoholic beverage sales generated in a foodservice operation.

When using the USAR, the physical layout of the income statement for different types of businesses can vary somewhat in their formats. For example, a foodservice operation that serves alcohol would want its income statement to identify the revenue generation and costs associated with serving alcoholic beverages. In a family-style pancake restaurant that serves only food and nonalcoholic beverages, the income statement would not include a line for alcoholic beverages.

The USAR's suggested income statement format recommends, at minimum, that food and (alcoholic) beverage sales be reported separately. There are a variety of reasons for this recommendation. These include the ability to better control costs when the sales are separated, and the requirement of most states to record alcohol sales separately from food sales.

In Figure 2.3, Jerry's food sales are separated from alcoholic beverage sales. In other operations, individual revenue categories may be created for catering sales, on-premises banquet sales, take-out versus dine-in sales, merchandise (e.g. logoed hats, cups, and T-shirts), or any other revenue category that would be helpful to the operation's managers as they analyze their sales.

In some foodservice operations, the sale of logoed items and other merchandise types can be significant. Readers familiar with the Cracker Barrel restaurant group are likely aware of the significant amount of merchandise it sells. Note: Even if an operation's merchandise sales are relatively small, they should be reported separately on the income statement.

What Would You Do? 2.1

"So, which is it?" asked Laszlo, the dining room manager at the 200-seat Harvest House Restaurant, "were our sales up or down last month compared to the previous year?"

"Both" replied Jessica, the manager of Harvest House.

"How can they be up and down at the same time?" asked Laszlo.

"Well," replied Jessica, "our overall topline sales last month were about the same as the prior year, but when I looked into it closely and checked the POS system, our actual dine-in sales were lower, and our carry-out sales were higher."

Assume you were the owner of Harvest House. How important would it be for you to know the proportion of your operation's sales that came from dine-in versus carry-out business? Would you recommend that the USAR income statement prepared for your operation be designed to clearly show these two food revenue categories separately? Why or why not?

Total Cost of Sales

Food and beverage **cost of sales** (Line 5) are entered separately on the income statement with the cost of food (Line 6) incurred to create food sales reported separately from the cost of (alcoholic) beverages (Line 7) incurred to create beverage sales. Without this separation, an inefficient

Key Term

Cost of sales: The total cost of the products used to make the menu items sold by a foodservice operation.

food operation could be covered up by a highly profitable alcoholic beverage operation (or the reverse could occur).

An operation's Total Cost of Sales (Line 8) reported on the income statement is the sum of Lines 6 and 7.

Total Cost of Sales is an operation's actual food and beverage product costs after considering inventory on hand at the beginning and end of the accounting period, employee meals, and other factors.

In many retail businesses, "cost of sales" is calculated as:

Beginning Inventory + Purchases − Ending Inventory = Total Cost of Sales

This simple formula is not acceptable for most foodservice operations because with it the expense of any employee meals provided would be classified as part of the Total Cost of Sales. If these meal costs are incurred, they would be better accounted for as part of an operation's Employee Benefits (Line 12).

In addition, when operators separate their costs into food costs and beverage costs, the value of food products transferred from the kitchen to the bar (e.g. fruits, juices, and garnishes used to make drinks) must be accounted for as a beverage expense. Also, properly accounting for the value of products transferred from the bar to the kitchen (e.g. wine and beers used for cooking) will provide a more accurate value for the actual cost of sales for food products.

A better model to calculate cost of sales for food in many foodservice operations is:

Beginning inventory (food)	_____
Plus:	
Food purchased during the accounting period	_____
Equals:	
Food available for sale	_____
Less:	
Ending inventory (food)	_____
Equals:	
Cost of food consumed	_____
Less:	
Employee meals	_____
Transfers to beverage cost	_____
Plus:	
Transfers from beverage cost	_____
Equals:	
Cost of sales (food)	==========

The same formula can be used to calculate cost of sales for beverages, with the exception that there will not be a reduction for employee meals.

Total Labor

While **payroll** generally refers to salaries and wages a foodservice operation pays to its employees, the USAR income statement provides greater labor cost-related details.

In Figure 2.3, an operation's total payroll cost is reported under the heading "Labor" (Line 9), and it is separated into three categories:

Key Term

Payroll: The term commonly used to indicate the amount spent for labor in a foodservice operation. Used for example in "Last month our total payroll was $28,000."

✓ Management
✓ Staff
✓ Employee Benefits

Management (Line 10) includes the total amount of salaries paid during the accounting period, and Staff (Line 11) refers to payments made to hourly (non-salaried) workers. The employee benefits category (Line 12) includes the cost of all benefits payments made for managers and hourly workers. Some benefit payments are mandatory (such as FICA [Social Security]) payments, while others are voluntary (e.g. the cost of providing employee health insurance).

Total labor expense (Line 13) is the sum of Lines 10, 11, and 12, and is listed prominently on a USAR income statement because controlling

Key Term

Total labor: The cost of the management, staff, and employee benefits expense required to operate a business.

and evaluating total labor cost is critical in every foodservice operation. In fact, many operators feel it is even more important to control labor costs than product costs. One reason is because, in some operations labor payments and labor-related benefit costs comprise a larger portion of their operating costs than do food and beverage product costs.

Prime Cost

Prime cost (Line 14) is calculated as an operation's total cost of sales (Line 8) added to its total labor cost (Line 13). Prime cost is clearly listed on the income statement because it is an excellent indicator of an operator's ability to control

Key Term

Prime cost: An operation's cost of sales plus its total labor costs.

product costs (cost of sales) and labor costs, the two largest expenses in most foodservice operations. The prime cost concept is also important because, when prime costs are excessively high, it is difficult to generate a sufficient level of profit in a foodservice operation even when other controllable and noncontrollable expenses are maintained.

Other Controllable Expenses

As shown in Figure 2.3, after prime cost, the next major section of the income statement is used to identify "Other Controllable Expenses" (Line 15). There are two major issues that concern managerial accountants as they consider expense data to be included on a USAR income statement. These relate to the **timing** and the **classification** (placement) of the expense.

Just as a foodservice operator must record all revenue generated during the accounting period that is addressed by the income statement, so too must the operator make sure all related expenses

Key Term

Timing (expense): Determining when to place an expense on an income statement.

Key Term

Classification (expense): Determining where to place an expense on an income statement.

incurred have been reported. If not done properly, expenses would be understated, and profits would be overstated. Alternatively, to intentionally (or unintentionally) overstate expenses and include expenses not incurred would understate profits.

To illustrate, assume a foodservice operator produces a monthly income statement. They are now preparing an income statement for January, the first month of the operation's fiscal year. The operator pays a property tax bill twice per year, and one-half of the bill is due in February, and the remaining balance is due six months later (August). For the January income statement now being produced, should the operator enter:

✓ 1/12 of the annual bill?
✓ 31/365 of the annual bill?
✓ "$0.00," because no property tax was paid in the month of January?

While the most important task of the operator is to ensure they record all expenses, and several approaches "could" be used, the best approach in this scenario would be to record 1/12 of the annual bill each month, as that approach best represents the actual monthly "cost" of the operator's property taxes expense.

As expense timing and classification decisions are made, the amounts of these **Other Controllable Expenses** are listed on the income statement. In general, these are expenses that can be influenced by a foodservice operator's own decisions. The USAR identifies and recommends the use of several "other controllable expenses" categories.

Key Term

Other Controllable Expenses: Expenses that a foodservice operator can influence with increases or decreases based on business decisions. Examples include marketing costs and utility costs.

The specific other controllable expense categories of an operation listed on its income statement can vary. As shown in Figure 2.3, commonly reported controllable expenses include:

Direct Operating Expenses (Line 16): In this expense category, operators list the cost of uniforms, laundry, supplies, menus, kitchen tools, and other items incidental to service in the dining areas that provide support in the kitchen and storage areas.

Music and Entertainment (Line 17): These costs, if significant, should be shown separately. Many foodservice operations offer little or no music or entertainment. When this is the case and expenses are small, they may be recorded as "Miscellaneous" within the direct operating expense category (Line 16).

Marketing (Line 18): This expense category includes newspaper, magazine, radio and TV, and Internet advertising, as well as other expenses for outdoor signs and direct mailings. Loyalty program costs, donations, and special events that promote the business can be additional marketing costs. In some cases, marketing fees associated with third-party meal delivery companies such as DoorDash and Uber Eats are also included in this category (see Chapter 5).

Utilities (Line 19): This category includes the cost of water, sewage, electricity, and gas used to heat and/or cool a building. When facilities are rented and the restaurant pays the utility bills, these costs are recorded as "utilities" rather than rent.

General and Administrative Expenses (Line 20): This group of costs includes expenses generally classified as operating "overhead." These expenses are necessary for the operation of the business as opposed to being directly connected with serving guests. This category most often includes the cost of items including office supplies, postage, credit card fees, telephone charges, data processing, general insurance, professional fees, and security services.

Repairs and Maintenance (Line 21): These expenses include painting and decorating costs, maintenance contracts for elevators and machines, and repairs to an operation's various equipment and mechanical systems. It is not used to record the purchase of new equipment.

The individual entries in the Other Controllable Expenses category (Lines 16–21) are summed to determine the Total Other Controllable Expenses amount ($343,611 on Line 22). That amount and Prime Cost ($1,547,399 on Line 14), are subtracted from Total Sales ($2,306,110) to yield the operation's **Controllable Income** ($415,100 on Line 23).

The Controllable Income line amount is often used to evaluate the effectiveness of foodservice operators because it represents sales minus only those costs and expenses that managers generally can directly control or influence. In many foodservice companies, the amount of an operation's controllable income is the basis for determining at least some portion of a foodservice manager's incentive or bonus pay.

Key Term

Controllable income: In a USAR-formatted income statement, the amount of money remaining after a foodservice operation's Prime Cost (Line 14) and Total Other Controllable Expenses (Line 22) are subtracted from Total Sales (Line 4).

Key Term

Noncontrollable expenses: Costs which, in the short run, cannot be avoided or altered by management decisions. Examples include lease payments and depreciation.

Noncontrollable Expenses

Noncontrollable expenses (Line 24) are listed next on the income statement, and they include all costs *not* under the immediate control of management.

Noncontrollable costs are often referred to as **fixed costs**. Unlike **variable costs** (those that vary with the amount of sales such as staff wages), fixed costs remain unchanged regardless of an operation's sales volume.

Noncontrollable expenses include those previously committed and which must be incurred regardless of an operation's sales volume. In a foodservice operation, examples of noncontrollable expenses include rental payments, interest on long-term loans, **depreciation**, and **amortization** charges.

Noncontrollable expenses listed in Figure 2.3 include:

Occupancy costs (Line 25): These include the cost of renting buildings and land, property taxes, and insurance on **fixed assets**. These expenses can vary considerably between foodservice operations. Since the owners of a business most often control these costs, they are only infrequently under the direct control of those operating the business.

Equipment leases (Line 26): These expenses include the costs incurred from leasing or renting equipment used in a foodservice operation. Common examples include charges for the leasing of POS systems, beverage dispensing equipment, and ice machines.

Depreciation and Amortization (Line 27): These **noncash expenses** are the result of depreciating buildings and **furniture, fixtures, and equipment (FF&E)** . In addition, this expense includes the amortization of leaseholds (the accounting term for an asset that is being leased) and any leasehold improvements.

To illustrate the profit variability caused by noncontrollable expenses, assume two foodservice operators manage almost identical units. Both units generate about $1 million in sales and

Key Term

Fixed cost: An expense that remains constant despite increases or decreases in sales volume.

Key Term

Variable cost: An expense that generally increases as sales volume increases and decreases as sales volume decreases.

Key Term

Depreciation: The allocation of the cost of equipment and other tangible assets based on the projected length of their useful lives.

Key Term

Amortization: The practice of spreading an intangible asset's cost over that asset's useful life.

Key Term

Occupancy costs: Costs related to occupying a space including rent, real estate taxes, personal property taxes, and insurance on a building and its contents.

Key Term

Fixed asset: An asset such as land, building, furniture, and equipment that is purchased for long-term use and is not likely to be converted quickly into cash.

both restaurants have $250,000 in controllable income. One operation has total noncontrollable expense of $100,000 and thus generates a **Restaurant Operating Income** (Line 29) of $150,000 ($250,000 controllable income − $100,000 noncontrollable expenses = $150,000 restaurant operating income).

The second operation has a total noncontrollable expense of $200,000 and thus generates a restaurant operating income of $50,000 ($250,000 controllable income − $200,000 noncontrollable expenses = $50,000 in restaurant operating income). In this example, it is likely that both operators are doing good jobs managing their businesses even though the restaurant operating income of the first unit is three times that of the second unit.

The individual entries in the Noncontrollable Expenses section of the income statement are summed to yield Total Noncontrollable Expenses (Line 28). The amount in Line 28 is then subtracted from Controllable Income (Line 23) to yield the operation's Restaurant Operating Income (Line 29).

Key Term

Noncash expense: Expenses recorded on the income statement that do not involve an actual cash transaction. Examples include depreciation and amortization expenses where an income statement charge reduces operating income without a cash payment.

Key Term

Furniture, fixtures, and equipment (FF&E): Movable furniture, fixtures, or other equipment that have no permanent connection to a building's structure.

Key Term

Restaurant operating income: All an operation's revenue minus all of its controllable and noncontrollable expenses.

Find Out More

Depreciation is a noncash expense that is calculated and then allocated over the time an asset is useful to a business. Businesses can take advantage of depreciation rules by deducting from taxable income the expense of using a portion of the value of their assets according to a predetermined schedule.

The U.S. Internal Revenue Service (IRS) establishes the rules for allowable depreciation amounts. There are several alternative depreciation methods available to foodservice operators, and the specific methods used to calculate allowable depreciation can vary based on the type of asset being depreciated and the operator's preference.

Appropriate depreciation method calculations can be complex, but the IRS provides specific guidance to those who are calculating depreciation amounts. To find out more about depreciation methods that are legally allowable, enter "IRS rules on how to depreciate property" in your favorite search engine and review the results.

Profits (Income)

It is interesting that the word "profit" does not actually appear anywhere on a USAR-formatted income statement. Some foodservice operators consider "Restaurant Operating Income" (Line 29) to be their business's profit. The reason: it represents an operation's sales (revenue) minus all controllable and noncontrollable expenses, and it reflects the profit formula presented earlier in this chapter:

Revenue – Expenses = Profit

Restaurant Operating Income on Line 29 does not consider any **interest expense** payments made by a business. Interest Expense (Line 30) is the cost of borrowing money and is recorded on this line of the income statement.

Key Term

Interest expense: The cost of borrowing money.

Restaurant operating income may be further adjusted to account for corporate overhead, interest expense, or other owner-controlled expenses.

Income Before Income Taxes (Line 31) on the income statement is calculated as Restaurant Operating Income minus Interest Expense.

When an operation's revenue exceeds its expenses, the amount of the income before income taxes is shown on the income statement. However, if expenses *exceed* revenue, the result-

Key Term

Income before income taxes: The amount of money remaining after an operation's interest expense is subtracted from the amount of its restaurant operating income. Also, a business's profit before paying any income taxes due on the profits.

ing negative numbers (losses) on the income statement are typically designated in one of the three ways:

1) By a minus ("–") sign in front of the number. For example, a $1,000 loss would be presented on the statement as –$1,000.
2) By brackets "( )" around the number. For example, a $1,000 loss would be presented on the statement as ($1,000).
3) With red ink rather than black ink to designate the loss amount. For example, a $1,000 loss would be presented on the statement as "$1,000," but the number would be printed in red. Note: This approach gives rise to the slang phrase to "operate in the red:" a business that is not making a profit. In a similar vein, to "operate in the black" indicates the business is profitable.

In Figure 2.3, the operation's Income Before Income Taxes (Line 31) is a number that is frequently referred to as a business's "profit." Note that the sample income statement shown in Figure 2.3 does *not* include a line for income taxes due. If the business was operating as a corporation (a taxable entity), an additional expense line for income taxes would be included in the income statement.

Many foodservice operations, however, are operated as a subchapter S corporation (S corporation), limited liability companies (LLC), partnerships, or sole proprietorships. In these entities, taxable income or losses flow through the business and are included on the income tax returns of the owners, shareholders, members, and partners. For this reason, the income statements prepared for these nontaxable entities reflect income up to but not including income taxes due.

As a final note on income statements, it is important to recognize one key financial metric that is *not* listed on a USAR income statement: EBITDA (earnings before interest, taxes, depreciation, and amortization).

Most foodservice owners are interested in knowing their operation's ability to generate cash. An operation's EBITDA is a measure that business owners and operators can use to determine their net cash (before taxes) income.

Key Term

EBITDA: Short for "earnings before interest, taxes, depreciation, and amortization." EBITDA is used to track and compare the underlying profitability of a business regardless of its depreciation assumptions or financing choices.

EBITDA is a unique number because it does not include the non-cash operating expenses of interest, taxes, depreciation, and amortization. Payments for these items are listed on the income statement (or elsewhere on a foodservice operation's financial statements). However, they are not related to the day-to-day core operation of a business. For example, the interest paid on debts is listed on the income statement, but this expense reflects how the business was financed and not the ability of the business to generate sales or profits. Similarly, income taxes due may or may not be listed on an operation's income statement, but these do not affect how managers operate daily.

Calculating EBITDA from a USAR formatted income statement is an easy two-step process:

Step 1: Identify Restaurant Operating Income

In Figure 2.3, Restaurant Operating Income is listed as $253,590 (see Line 29).

Step 2: To Line 29, add back the amount listed for Depreciation and Amortization (see Line 27):

Restaurant Operating Income	$253,590
+ Depreciation and Amortization	$ 41,510
= EBITDA	$295,100

In this example, EBITDA for Jerry's Restaurant is $295,100.

Many foodservice operators view EBITDA as a good way to assess the earning power of their businesses and to compare their own operations with similar operations having different debt levels or depreciation amounts.

However, it is also important to recognize important limitations of EBITDA. For example, interest and income taxes are actual business expenses that must be paid by the foodservice operation. In addition, depreciation and amortization should reflect the decline in the real value of a business's assets. Each of these may be "noncash" expenses, but the expenses are real! In general, however, most owners would agree that EBITDA is one good way to help assess the cash-generating ability of a foodservice operation.

Break-even Point Analysis

While the income statement documents the past performance of a foodservice operation, most operators would like to know the **break-even point** of their businesses on a daily, weekly, or monthly basis.

Key Term

Break-even point: The sales point at which total cost and total revenue are equal, and there is no loss or gain for a business.

In effect, by determining the break-even point, the operator is answering the question, "How much sales volume must be generated before I begin to make a profit?" Beyond the break-even point, the operator wants answers to another question: "How many sales dollars and business volume must I generate to achieve make my *target* profit level?"

Experienced foodservice operators know that some accounting periods can be more profitable such as those in businesses that experience "busy" and "slow" periods. For example, a ski resort may experience significant food sales during the ski season but then have a greatly reduced volume (or may even close) during the summer. Similarly, a country club manager in the Midwestern United States knows revenue from greens fees, golf outings, and in-house food and beverage sales can be higher in the summer months. However, they also know the golf course will likely be closed for several winter months.

Experienced foodservice operators who understand fixed and variable costs know that costs as a percentage of sales are generally reduced when sales are high, and these percentages increase when the sales volume is lower. The result, in most cases, is greater profits during high-volume periods and lesser profits in lower volume periods. This relationship between volume, costs, and profits is easier to understand when examined graphically as shown in Figure 2.4.

The x (horizontal) axis in Figure 2.4 represents sales volume: the number of covers (guests) served or dollar volume of sales that were generated. The y (vertical) axis represents the costs associated with generating these sales. The Total Revenues line starts at 0 because, if no guests are served, no revenues are generated. The Total Costs line starts further up the y axis because fixed costs are incurred even if no sales are made.

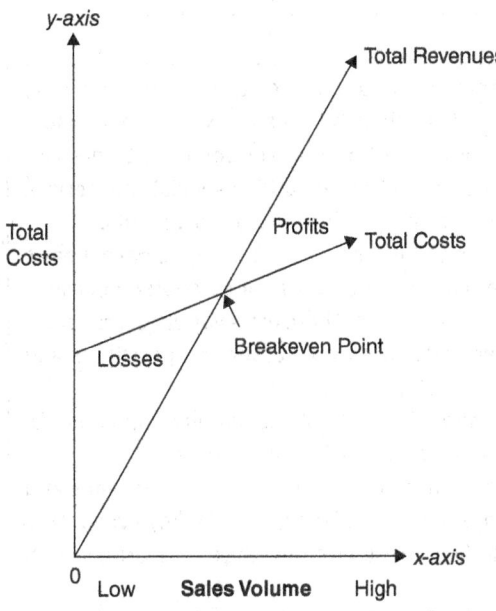

Figure 2.4 Cost/Volume/Profit (Break-even)

The point at which the two lines cross is the break-even point at which operating expenses exactly equal revenue. Stated another way, when sales volume equals the sum of its total fixed and variable costs, the break-even point has been reached. Below the break-even point, costs are higher than revenues and losses occur. Above the break-even point, revenues exceed the sum of the fixed and variable costs required to make the sales, so profits are generated.

Technology at Work

Every foodservice operator should know the break-even point of their business. Simply stated, a break-even point analysis is a calculation that tells foodservice operators what quantity of menu items must be sold to be profitable. Also, it helps them produce a pricing strategy that will not only cover their costs but will also help ensure that they make a profit. The break-even point for every foodservice operation will be different, but it can be calculated if operators know their fixed and variable costs.

Managerial accountants can use several different accounting methods and software packages to calculate a break-even point for a business. To see some different mathematical approaches to the calculation of a break-even point, go to the YouTube site. When you arrive, enter "how to calculate a restaurant break-even point" in the search bar and view one or more of the posted videos.

Find Out More

Some foodservice business owners and operators take personal responsibility for preparing the accounting records of their businesses, while others use a certified public accountant (CPA) or accounting services company to prepare their financial reports and statements. Still others utilize special foodservice accounting software packages that prepare their financial documents.

In all cases, however, it is important that foodservice managers have a basic understanding of how results are reported and how finance-related decisions are made in a successful foodservice operation. There are a variety of resources available to help operators better understand the accounting principles and practices used in the foodservice industry.

One of the best, and most detailed, is the book, *Accounting and Financial Management in Foodservice Operations*, published by John Wiley.

To learn more about the content and availability of this up-to-date and extremely valuable accounting information resource, enter "Wiley Accounting and Financial Management in Foodservice Operations" in your favorite search engine and review the results.

This chapter has addressed the ways foodservice operators properly account for their revenue and expenses. For most operators, one of the most important factors that impact their revenue generating ability is the quality of their marketing programs. Regardless of its type and size, the effective marketing of a foodservice operation is important for its short-and long-term success, and this will be the sole topic of the next chapter.

What Would You Do? 2.2

"I can buy it at a great price," said Dawn Fleming who was talking about the Wayside Restaurant. The property was for sale, and Dawn, a real estate investor, was meeting with Lynette Nystrom, her friend and experienced restaurant manager. "As I read its income statement, it's losing about 5 cents on each dollar of sales now," continued Dawn, "but I know we can turn that around!"

Lynette considered Dawn's proposal that they form a partnership, acquire the restaurant, and share the profits they will make. She knew that, before it was possible to share profits, they would have to make a profit. That meant to go from losing 5 cents per dollar to making money, they would have to increase sales and/or reduce costs, and she pointed that out to Dawn.

"Well," Dawn replied, "I'm not sure we need to increase the sales at all. If we buy it at the right price, I think we just need to reduce the operating costs, and you can do that, I know you can!"

Assume you were Lynette. Why would it be important for you to know about this operation's fixed and noncontrollable costs under the arrangement proposed by Dawn? Why would it be important to know about the operation's likely variable and controllable costs and its break-even point if you agreed to operate it?

Key Terms

Accounting
Accountant
Guest check average
Managerial accounting
Bookkeeping
Point-of-sale
   (POS) system
Accounting period
Income statement
Balance sheet
Statement of Cash
   Flows (SCF)
Positive cash flow
Uniform System of
   Accounts for
   Restaurants (USAR)
Profit center
Cost center

Return on investment
   (ROI)
28-day accounting
   period
Topline revenue
Total sales
Cost of sales
Payroll
Total labor
Prime cost
Timing (expense)
Classification
   (expense)
Other Controllable
   Expenses
Controllable Income
Noncontrollable
   expenses

Fixed cost
Variable cost
Depreciation
Amortization
Occupancy costs
Fixed asset
Noncash
   expense
Furniture, fixtures,
   and equipment
   (FF&E)
Restaurant operating
   income
Interest expense
Income before
   income taxes
EBITDA
Break-even point

Operator's 10-Point Tactics for Success Checklist

Evaluate your need for, and the current status of, each of the following operational tactics. For those tactics you think are important, but not yet in place, develop an action plan for its implementation including who will be responsible for the tactic's completion and the target date by which it should be completed.

Tactic	Don't Agree (Not Done)	Agree (Done)	Agree (Not Done)	If Not Done	
				Who Is Responsible?	Target Completion Date
1) Operator recognizes the importance of utilizing professional accounting methods to report the operating income and expenses of a business.	____	____	____		
2) Operator can state the differences between bookkeeping, summary accounting, and financial analysis.	____	____	____		
3) Operator can explain the purpose of preparing an accurate income statement, balance sheet, and statement of cash flows (SCF) for a business.	____	____	____		
4) Operator can identify and explain the five-step process used to analyze the financial performance of a business.	____	____	____		
5) Operator understands the purpose of the Uniform System of Accounts for Restaurants (USAR) and the advantages of using it in the preparation of financial reports.	____	____	____		

				If Not Done	
Tactic	Don't Agree (Not Done)	Agree (Done)	Agree (Not Done)	Who Is Responsible?	Target Completion Date
6) Operator can identify the information to be placed in the "Sales" portion of an income statement prepared using the USAR-recommended format.	___	___	___		
7) Operator can identify the information to be placed in the "Expense" portion of an income statement prepared using the USAR-recommended format.	___	___	___		
8) Operator can identify the information to be placed in the "Restaurant Operating Income" (Profit) portion of an income statement prepared using the USAR-recommended format.	___	___	___		
9) Operator can state the importance of calculating EBITDA as a measure of a business's ability to generate cash.	___	___	___		
10) Operator can state the purpose of conducting a break-even point analysis.	___	___	___		

3

Successful Marketing in Foodservice Operations

What You Will Learn

1) The Four Ps of Marketing
2) How to Identify a Target Market
3) How to Create and Implement a Marketing Plan
4) How to Evaluate Marketing Plan Results

Operator's Brief

In this chapter, you will learn that marketing is the process by which you as a foodservice operator communicate important product and services information to your current and potential guests. The major goal of successful marketing is to grow the number of guests who will choose your operation from among their alternatives.

Many foodservice operators find it convenient to focus on the 4Ps of marketing as they craft specific marketing messages to deliver to potential guests. The 4Ps are:

1) Product
2) Place
3) Price
4) Promotion

"Product" includes all the menu items and services you offer to potential guests. "Place" is where and how your guests can buy from you. "Price" is the amount guests pay for the products and services they select, and it includes how they pay for them.

"Promotion" has a number of key components. These are advertising, personal selling, promotions, publicity, and public relations. In this chapter, you will learn about the roles of all these important components of promotion.

Experienced foodservice operators know that their goal is not to try to appeal to all possible customers, but rather to identify and direct their marketing efforts to target markets: the guests who will be *most likely* to buy from them. In this chapter, you will learn why your target markets must be identified, how to do so, and procedures to determine the needs of these markets.

After you have determined the marketing messages you wish to send and have identified those to whom you wish to send the messages, you must create a marketing plan; a written plan detailing your operation's marketing efforts for a specific time period. The marketing plan will provide details about marketing activities to be undertaken, and it states what is to be done, when it is to be done, and who will do it.

After a marketing plan has been implemented, its results must be evaluated. In this chapter, you will learn how you can assess the effectiveness of your marketing plan so it can be repeated in the future and/or can be modified and improved.

CHAPTER OUTLINE

The Importance of Successful Marketing
The 4 Ps of Marketing
     Product
     Place
     Price
     Promotion
The Importance of Target Markets
     Why Target Markets Must Be Identified
     How Target Markets Can Be Identified
     How to Identify the Needs of a Target Market
Marketing Plans
     Creating the Marketing Plan
     Implementing the Marketing Plan
     Evaluating Marketing Plan Results

The Importance of Successful Marketing

All foodservice operations must successfully serve their customers if they are to stay in business. Whether a foodservice operation's customers are called guests, students, patients, or any other title, satisfying those being served is essential for the continued operation of any successful foodservice. In fact, foodservice operators should consider their main job to be adding new customers while retaining current customers.

To ensure a steady flow of customers, foodservice operators must understand **marketing**: all the ways a foodservice operation communicates with current and potential guests.

Marketing involves how a business informs potential guests about what it is offering for sale and, most importantly, why customers should buy from them. The major goal, or purpose, of effective marketing is to obtain and retain a *growing* base of satisfied guests.

Key Term

Marketing: The varied activities and methods used to communicate a business's products and service offerings to its current and potential guests.

The 4 Ps of Marketing

With so much essential information to be shared with current and potential guests, many foodservice operators find it convenient to focus on the **4 Ps of marketing** to increase their success when promoting their business's products and services.

The 4 Ps of marketing are illustrated in Figure 3.1.

The actual use of the 4 Ps of marketing is sometimes referred to as an operation's **marketing mix** because different foodservice operators emphasize each of these four components in different ways as they market their businesses.

Key Term

4 Ps of Marketing: A way to categorize a business's marketing strategy based on the products sold, places where they are sold, the prices at which they are sold, and the promotional efforts used to sell them.

Key Term

Marketing mix: The specific ways a business utilizes the 4 Ps of marketing to communicate with its potential customers.

Product

In a foodservice operation, the specific menu items sold and how they are sold are the operation's products. Most foodservice operators spend time thinking about and analyzing the products they sell. The reason: they know they must consistently satisfy their guests' product needs. Consequently, most foodservice operators are proud of the menu items they offer and have spent significant time developing the recipes and cooking methods used to create these products.

Products offered for sale are important to guests, but the products sold by foodservice operators varies widely from one operation to the next. For example, it is well-known that hamburgers are the single most popular menu item sold in the United States. In some operations, the production of a quality hamburger may

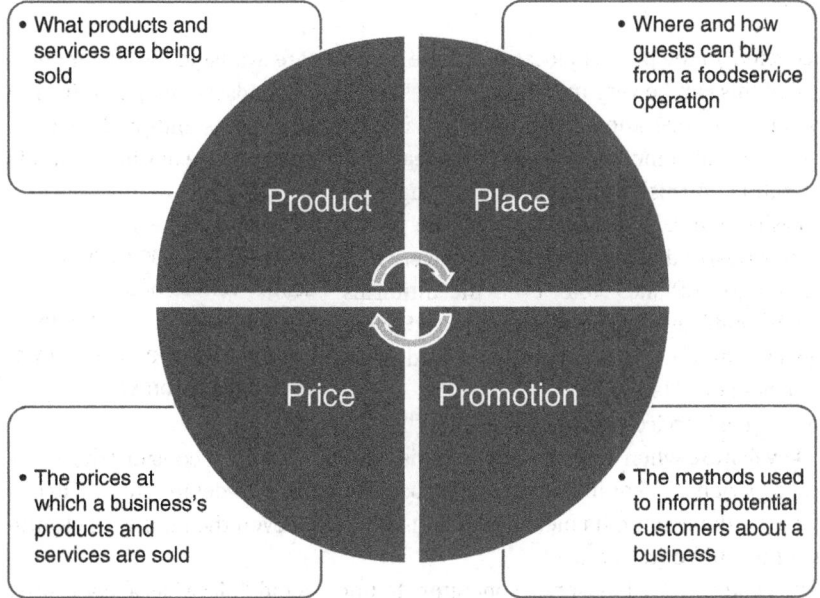

- What products and services are being sold

Product

- Where and how guests can buy from a foodservice operation

Place

Price

- The prices at which a business's products and services are sold

Promotion

- The methods used to inform potential customers about a business

Figure 3.1 The 4 Ps of Marketing

involve cooking a 2-ounce (purchase weight) beef patty and then placing it on a bun with ketchup, mustard, and a single pickle slice. The hamburger may then be wrapped and, when sold, it may be placed in a paper sack for delivery to a customer in the operation's drive-through lane.

In a different operation, a hamburger may be made by cooking to medium rare a 6-ounce (as purchased (AP) weight) patty of extremely high-quality beef. The patty may then be topped with cheddar cheese, leaf lettuce, and a slice of red onion. The hamburger is served on a brioche bun, and the product may then be picked up in the kitchen by a server who delivers the hamburger to the guest's table.

In this example, both operations are offering a "hamburger," but the products received by their guests will be very different. The prices paid by those guests will likely be very different as well. In foodservice operations, a menu item may vary by portion size, method of preparation, speed of delivery, and packaging method to name just a few product variations.

For many foodservice operators, a strong marketing emphasis focused on the actual products that are sold makes good sense. This is especially so when the products are especially unique. Examples include foodservice operations serving vegetarian foods, specific ethnic foods such as Thai or Vietnamese, or those whose entire operation emphasizes a particular menu category such as steaks or seafood.

Place

Place refers to the physical location where products are available for sale or delivery, and this can be very important to the success of a foodservice operation. The location of an operation may include its indoor dining areas and outdoor patio areas, and it also includes the **curb appeal** of the operation's exterior, including its lighting, furnishings, and exterior signage. Increasingly, meals are picked up by the guests or a third-party delivery person for delivery and consumption off-site. Note: even the uniforms worn by staff and the background music being played within the operation may be considered a component of "Place."

Key Term

Curb appeal: The general attractiveness of a foodservice operation when viewed from the outside by a potential customer.

When foodservice operations market "Place" as a key feature when appealing to potential guests, they must consider the entire guest experience. That includes paying special attention to details such as décor, the size of the type on its menus or menu-board, and even the quality of cleanliness of its public restrooms!

One good way for foodservice operators to understand "Place" as a key feature in the marketing message is to see the operation from their guests' viewpoints. What do they see, hear, and smell as they arrive at a foodservice operation's entrance area or drive-through lane? The answers to these questions are very important, and the foodservice operator must ensure these factors contribute positively to the image of "Place."

Convenient, well-lighted parking areas, clean entrance and waiting areas, and tidy dining spaces are important to all guests regardless of the menu items being served. It is also important to recognize that, while cleverly designed buildings and beautiful interiors may draw guests for initial visits, only the proper serving of quality food and beverage products will encourage those guests to return.

Foodservice operators who emphasize "Place" in their marketing mix often do so because their physical location can be a powerful customer draw. Examples include those operations located to offer a "beach view" or that are "right off the exit," or are "located in the heart of the theater district."

Price

Price is so important to the success of foodservice operations that it will be the single topic addressed in Chapter 7. In many cases, the price of a product or service directly influences its sales volume and, as a result, a foodservice operation's revenue and resulting profits.

Experienced foodservice operators know that the prices charged for the menu items they sell can be affected by numerous things. These include product and labor costs, overhead costs, the pricing trends of competitors and even, in the case of the sale of alcoholic beverages, government regulations.

From a marketing perspective, menu prices send a very clear message to potential guests about the value an operation provides in its products and services. When foodservice operators use higher quality ingredients or provide larger portion sizes, menu prices will reflect these issues. Similarly, foodservice operations located in highly desirable areas (e.g. those with ocean-view dining rooms or locations in popular and trendy downtown areas, prices may reflect the foodservice operation's location desirability as well).

Proper pricing is important because, if an operation's prices are set too low, the operation may be popular with guests, but it may also have difficulty achieving its financial goals. Similarly, if prices are set too high, the operation may attract too few customers. This may also cause the operation to have trouble achieving its financial goals.

As a part of its marketing function, "Price" also includes how guests actually pay for their purchases. Historically, foodservice operations accepted cash, personal checks, and credit and debit cards as payment. Today, few operations accept personal checks but, increasingly, they offer their guests the opportunity to pay through payment apps such as Apple Pay, Google Pay, or Samsung Pay. When guests are offered these payment options, that price-related information must also be properly communicated to them.

For many foodservice operators, an emphasis on price means promoting their "low-priced" menu items. While that can be an effective strategy, foodservice operators should also recognize that advertising the "lowest price" for its menu items will appeal to some guests, but it *may not* appeal to others.

Promotion

Promotion is the portion of an operation's marketing efforts that communicate directly with customers. Key components of promotion include:

Advertising
Personal selling
Promotions
Publicity
Public relations

Advertising is an example of a food operation's direct messaging to guests. Its key components include developing an effective advertising message and selecting the best tools (e.g. radio, print, television, text, or the Internet) to deliver that marketing message. Operators must also consider the frequency and costs of their message delivery.

Personal selling efforts in foodservice operations include efforts used on-premises and off-site. On-premises selling efforts can include signage about a featured item and recommendations by waitstaff and bartenders as they serve guests. Off-site selling efforts are those undertaken by an operation's owners, managers, or sales staff to increase local business. Examples may include developing websites, engaging in social media activity, and participation in local community events.

"Promotion" is one of the 4 Ps of marketing, but the same term is also used to identify specific offers that appeal to all customers and, in many cases, provide special rewards for loyal customers. For example, a foodservice operation may offer reduced pricing on a selected menu item during certain times of the day or week as a special promotion.

Publicity refers to communication about a business that has been placed in the media without the business paying for it directly. Publicity is important to a foodservice operation for two reasons. First, the operation benefits from positive publicity since it reflects well on the business. For example, a foodservice operation may sponsor a youth sports team. If that sponsorship is reported in a local newspaper, or on television, radio, or a popular website, the publicity will likely be positive.

A second reason that publicity is important is that not all publicity is positive. For example, in many communities, sanitation inspection scores assigned by the local health department for foodservice operations are made public in the newspaper or on television (Note: This topic was addressed at length in Chapter 1). If an operation's inspection score is not good, the result can be negative publicity. Similarly, a significant outbreak of a foodborne illness is likely to be widely reported in the local community. The publicity surrounding such an outbreak can have an extremely negative impact on an operation's reputation and its future success.

Today, a large number of popular websites allow guests to post comments about their most recent visits to a foodservice operation. These reviews will no doubt be widely read by potential guests. While many posted comments may be positive, some may be negative. Proactively responding to these comments posted on reviewer sites is an important topic and it will be addressed in detail in Chapter 5.

Finally, foodservice operations must employ public relations to promote and manage a positive public image: how the operation is viewed by the public.

An effective marketing strategy for a foodservice operation involves identifying the right customers and then using all the above promotion tools to reach potential guests with a compelling sales message.

Figure 3.2 summarizes the activities and goals of advertising, personal selling, promotions, publicity, and public relations.

Activity	Goal
Advertising	Delivering the marketing message in a cost-effective manner
Personal selling	Increasing customer counts, operational visibility, and guest check average through face-to-face interactions with customers
Promotions	Providing all guests (and especially loyal customers!) with special buying opportunities or rewards
Publicity	Increasing attention from the media at no cost to the operation
Public relations	Increasing name recognition and awareness of an operation in its own business community

Figure 3.2 Marketing Activities and Goals Summary

Technology at Work

Many foodservice operations use guest loyalty programs as an integral part of their marketing strategy. Essentially, these programs offer guests who already enjoy a foodservice operation's products and services some additional incentives to return often.

Typically, a customer loyalty program offers guests a discount, a free menu item, or specific merchandise as a reward for their frequent buying of qualifying purchases.

In the early days of foodservice rewards programs, customers were typically given a punch card to record the number of times they visited or the total value of their purchases over time. Today, specially designed customer loyalty program software does away with the need for such cards. Since they are easy to initiate and maintain today, nearly every commercial foodservice operator should utilize a customer loyalty program.

To review specific guest loyalty program software and apps appropriate for use in a foodservice operation, enter "restaurant customer loyalty program software" in your favorite search engine and view the results.

The Importance of Target Markets

Key Term

Some operators think their customer **market** consists of everyone who could possibly buy what they sell. While that may sometimes be

Market: The group of all people who could be customers of a business.

true, most foodservice operations know that their **target market** is the one that is of most importance to them.

Markets consist of all consumers who *could* buy what an operation sells, but target markets are important because they are the buyers who are *most likely* to become customers. As shown in Figure 3.3, a target market is a sub-segment of all the consumers who buy products from foodservice operators and will most likely want or need a specific foodservice operator's products or services.

Key Term

Target market: The group of people with one or more shared characteristics that an operation has identified as the *most likely* customers for its products and services.

Why Target Markets Must Be Identified

Target market identification is essential to effective and efficient foodservice marketing. Operators do not want to waste their time and money communicating to consumers who are unlikely to be interested in what the operators sell. Instead, foodservice operators want to carefully tailor their marketing efforts to their target markets.

For example, a very upscale coffee shop located in a downtown's business area is open to serve everyone. This type of coffee shop, however, would likely consider its target market to be customers willing to pay higher prices for higher-quality coffee products. Not all potential customers are likely willing to do that, at least not every time they purchase coffee.

In this example, the coffee shop's total market may well consist of all persons who are working in or visiting downtown, and who are coffee drinkers. However, its *target* market is those individuals who are different from all other customers because they are willing to pay premium prices for premium coffee products.

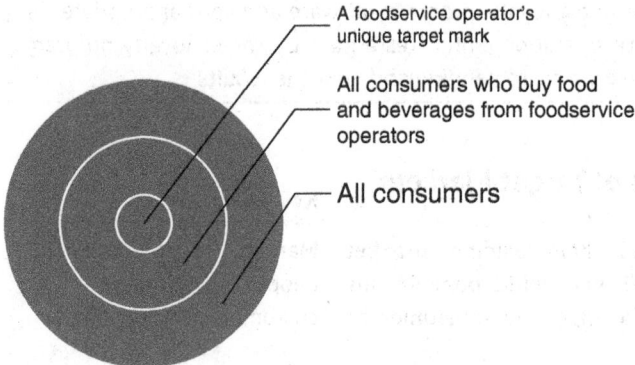

A foodservice operator's unique target mark

All consumers who buy food and beverages from foodservice operators

All consumers

Figure 3.3 Foodservice Operator's Unique Target Market

As another example, it is highly likely that a large number of guests might enjoy being served high-quality steaks or seafood in an elegant dining room. Foodservice operations serving these high-quality menu items have target markets that include guests with this desire. However, their true target market includes only the potential guests who have both a willingness to pay *and* the ability to pay the higher prices the operation will charge.

To be successful, foodservice operators must carefully identify their target markets and then work diligently to communicate how they will satisfy the wants and needs of that market. When they do, they will be successful in attracting new customers, retaining existing customers, and increasing their sales and profits.

How Target Markets Can Be Identified

Professional foodservice operators use the term, "target market," to describe the individuals and businesses that are most likely to buy their products and services. These are the customers with whom foodservice operators most want to communicate.

There are several ways to identify a foodservice operation's target market. One common method is to classify target markets by their shared characteristics, and there are three major groupings that are of most importance when doing this:

✓ Geographic
✓ Demographic
✓ Psychographic

Geographic

Target markets can be segmented geographically. Geographic segmentation means identifying a target market based on its location. This is a good strategy when a food operator serves customers in a particular area, or when a target market has different preferences based on where they are located. Target customers can be located by country, state, region, zip code, city, or even local neighborhood. Segmenting a target market based on its location in an urban or suburban area is another example of geographic segmentation.

Geographic segmentation is driven by the principle that people in a specific location may have similar needs, wants, and/or buying behaviors. For example, a pizza parlor operating near a major university may recognize that its target market includes primarily students who live in, or close to, the university campus. While that pizza parlor may serve nonstudents as well, it is likely that *most* customers will share their geographic location near the university. This common area will likely be more important than gender, income level, or race when the pizza parlor's best potential customers are identified. Recognition of target markets based on geography is especially important for advertising smaller independent foodservice operations. The reason: many potential customers will likely have some knowledge about the businesses being advertised because they will have seen the operation in their own neighborhoods.

Find Out More

Professional foodservice operators know that identifying a market by geography can do more than simply allow them to better target their advertising efforts. In some cases, a customer's geography will significantly influence the menu items to be offered.

For example, in Hawaii, many McDonald's operators offer Spam as a meat alternative with scrambled eggs, and Thailand McDonald's serves a samurai pork burger. In the United Kingdom, McDonald's serves mozzarella dippers and, in South Africa McDonald's serves the South African stack (two 100% British & Irish beef patties, Smoky BBQ sauce, Beechwood smoked bacon, Pepper Jack cheese, Hot & Spicy Mayo, and lettuce and a slice of tomato in a glazed paprika topped bun.) In China, you can buy a McDonald's Cilantro Sundae (vanilla soft serve ice cream covered with lime green sauce and then topped with a hefty pile of dried cilantro leaves).

To see more examples of restaurant operators tailoring their menus to the unique desires of guests found in various geographic locations enter "Chain restaurant menu variations by region" in your favorite search engine and view the results.

Demographic

Geographic segmentation separates guests based on *where* they are, while demographic segmentation separates guests based on *who* they are. Some foodservice operators find it best to classify their customers on the basis of demographics.

A target market's demographics consists of a list of statistical data about the characteristics of a specific group of people. In simple terms, a demographic is one characteristic of a larger population.

Examples of demographic characteristics of a potential target market include:

✓ Age	✓ Occupation	✓ Nationality
✓ Sex	✓ Education level	✓ Religion
✓ Marital status	✓ Income	
✓ Family size	✓ Race	

In some cases, a single demographic factor can completely include (or exclude) a potential group of customers. This is true, for example, for foodservice operators who sell alcoholic beverages in addition to food. In this example, only those who are of legal drinking age are potential customers for the operation's alcoholic beverage items. Note: a much wider range of customers with a different "age" demographic characteristic can be buyers of the operation's other menu items. For example, some foodservice operations offer children's menus only to customers younger than a specifically-identified age.

Other examples of directly addressing customers' demographic characteristics include foodservice operations that promote their kosher menu items (items prepared for those who follow Jewish dietary practices) and halal menu items (items prepared for those who follow Muslim dietary practices).

Psychographic

Psychographic segmentation is similar to demographic segmentation, but it addresses characteristics of a target market that are mental or emotional. Psychographic characteristics include customers' personality traits, attitudes, values, interests, and beliefs.

These attributes are not easily seen; however, they can give valuable insight into a target market's motives, needs, and preferences. While demographics provide factual characteristics of "who" the customers are, psychographics give foodservice operators insight into "why" their customers make their buying decisions.

One good example in the foodservice industry of appealing to a psychographic characteristic are operations that advertise themselves as being "eco-friendly." In this example, the operations seek to communicate their pro-environment business philosophy to a select group of people who feel it is important to adopt sustainable and **green practices**. These individuals could exist in several different geographic and demographic classifications.

Those foodservice operations providing "Early Bird" specials are another example of appealing to a psychographic characteristic. In this case, the customer characteristic of preferring to dine early in the evening is targeted by operators.

Key Term

Green practices: Those activities that lead to more environmentally friendly and ecologically responsible business decisions. Also known as "eco-friendly" or "earth-friendly" practices.

Food and beverage operations offering "Happy Hour" drink specials provide another example of utilizing a psychographic characteristic (customers seeking lower drink prices when foodservice operators sell the reduced-price drinks at specific times of the day).

Regardless of the segmentation approach used, the identification of target markets is critical to most foodservice operators. Some foodservice operations may even have multiple target markets. That is, they may have a primary target market (their focus), and a secondary target market (not as large but with great growth potential). In all cases, foodservice operators who understand their target markets can be in a better position to devise effective **marketing messages** geared directly toward the target markets.

How to Identify the Needs of a Target Market

After a foodservice operation has identified its target market, it must consider the specific needs, wants, and desires of that market. A customer's need can be defined as something they lack, or the difference between where the customer is and wants to be.

Key Term

Marketing message: How an operation communicates to its customers and highlights the value of the products and services that are offered to its target markets.

For example, a customer may be hungry (what they currently are), and they want to be well-fed (what they would like to be). This need for food causes them to seek a foodservice operation where they hope their need will be satisfied. If that customer's meal purchase meets their expectations, they would likely want to return to the operation in the future. If, however, they were not satisfied (the meal did not meet their expectations), they would not likely return in the future. It is important to remember that successful marketing involves first identifying and then meeting (exceeding) customers' expectations.

Hunger is one need that could generate a customers' desire for a foodservice operation, but it is not the only need. If foodservice operations existed only to meet hunger needs, they might all virtually look the same and offer the same menu items. However, foodservice guests have many more needs than just satisfying their hunger.

To illustrate, assume that a foodservice operation has determined that senior citizens are a major target market. While these customers certainly desire well-prepared menu items, they are also likely to desire a welcoming environment, more lighting, less noise, and easy-to-read (large type) menus. Their "want" list continues with comfortable seating, high levels of personal attention, and value for the meals they purchase. Each of these needs is in addition to their need to satisfy their hunger.

In a similar manner, cost-conscious customers utilizing the drive-through lane of a quick-service restaurant certainly desire well-prepared menu items. However, they are also likely to be interested in their order's speed of delivery, accuracy in filling their orders, and appropriate packaging of their selected to-go menu items. Each of these desires will be in addition to the need to satisfy their hunger at a relatively low price.

After target markets are identified, the needs and desires of these specific markets can also be defined. Foodservice operators use several ways to identify their target markets' needs. Three of the most important of these are to:

1) Listen to customers
2) Ask questions
3) Keep good sales records

Listen to Customers

Most foodservice operators understand the importance of interacting directly with their guests. These interactions are most helpful, however, when operators carefully listen to what their guests are saying. In most cases, customers will not hesitate to tell what they like and dislike about an operation's prices, menu items, and service levels.

It is important for foodservice operators to listen to their current customers for good and bad comments. If there are menu items or service procedures causing dissatisfaction among a target market, operators should be aware of this so they can make improvements.

One common way to discover what guests think about an operation is to circulate in dining areas and directly engage guests in conversation. Today, increasingly, foodservice operators find out what their customers are saying by reading the information left on on-line review sites. This nontraditional way of listening is especially important because customers sometimes are more open and honest when they are posting a review online than when they are talking to a manager or server in person.

Find Out More

As drive-throughs continue to increase in popularity among foodservice customers, operators are increasingly concerned about monitoring the time it takes to process a drive-through order because most drive-through customers are in a hurry.

A study conducted by SeeLevel HX; a customer experience measurement company, found wait times for receiving a drive-thru order increased by more than 25 seconds from 2020 to 2021.[*]

The study classified the total time customers wait for an order from the moment they enter the drive-through to the moment they receive their order. The study found that in 2021, drive-through customers waited an average of 6 minutes, 22 seconds. Note: in 2020, it was 5 minutes, 57 seconds.

The research involved nearly 1,500 drive-thru visits between July and August 2021, to 10 major fast-food brands including McDonald's, Burger King, Taco Bell, Chick-fil-A, and Dunkin (formerly Dunkin Donuts).

Wise foodservice operators will no doubt continue to carefully monitor the average amount of time it takes to serve their customers using drive-throughs. To find up-to-date information about drive-through service times, enter "average foodservice drive-through speeds," in your favorite search engine and view the results.

[*]https://www.prnewswire.com/news-releases/seelevel-hx-21st-annual-drive-thru-study-uncovers-delays-and-inaccuracy-as-qsrs-struggle-with-labor-shortage-301383881.html retrieved July 24, 2023

Ask Questions

Even the best foodservice operators are not mind readers, and one way to find out about the needs and wants of target customers is to simply ask them. Asking guests about their needs and wants provides valuable information and communicates to a target market that management cares about guests and how they might be best served. Increased customer loyalty is often a positive by-product of conversations that demonstrate a business person is interested in their customers' opinions.

In addition to holding conversations with guests, an extremely popular way to gather information about their wants and needs is through surveys. These surveys need not be complex, and guests might be asked to fill them out when they are on-site or online.

Typical questions that can provide valuable information include:

✓ How did you find out about our operation?
✓ How often do you visit our operation?
✓ How did you place your order with us?
✓ How would you rate the speed of our service?
✓ What items would you like to see us add to our menu?
✓ How would you rate the quality of our food?
✓ How would you rate the quality of our service?
✓ Do you feel that you get good value for what you spend at our operation?
✓ How likely would you be to recommend our operation to your friends and family?
✓ What changes would you suggest to make our products and services even better?

The best foodservice operators not only survey their guests regularly, but they carefully record and analyze survey information. They pay particular attention to those areas in which many members of their target markets indicate the same positive or negative responses.

Keep Good Sales Records

It is important to keep good records about what the foodservice operation's customers buy. When satisfying the needs of an operation's target market, sales records about menu item sales including when the sales are occurring provide important insight into the purchase behavior of the target market.

In addition to keeping good records about menu item sales (what and when), it is increasingly important to keep good records about how items are being sold. For example, in the not-too-distant past, most foodservice operations sold most of their menu items to on-site diners. More recently, many foodservice operations sell increasing numbers of their items with drive-through, pick-up, or third-party delivery options. When these are offered, it is the best practice for foodservice operators to monitor menu item sales based on how the items are served or delivered to their guests.

Marketing Plans

The successful marketing of a foodservice operation requires a formal plan of action. A **marketing plan** is a document foodservice operators can develop and use to guide them in their marketing efforts. The purpose of a marketing plan is to identify strategies and tactics a foodservice operation can use to achieve its marketing goals. In most foodservice operations, the marketing plan should be prepared annually.

Marketing strategies are the broad and long-term goals a foodservice operation wants to achieve. These strategies may also be called marketing goals or objectives. **Marketing tactics** are the specific steps and actions an operation takes to achieve its marketing strategies. Each marketing strategy (goal) identified helps foodservice operators select the various marketing tactics they will need to employ to achieve it.

A marketing strategy describes *what* a foodservice operation wants to achieve, and its marketing tactics detail *how* the strategy will be achieved. Marketing strategies are first analyzed and then marketing tactics to achieve the strategies are considered.

For example, one marketing strategy for a foodservice business may be to increase its take-out sales in the coming year by 5%. Various marketing tactics to achieve that sales increase may include specific advertising and promotional efforts that are carefully selected and implemented to help attain the revenue increase goal.

Marketing a foodservice operation in today's competitive environment is too complex to be attempted without a formal marketing plan. For some foodservice operations, a critical strategy will be to obtain new customers. For others, it may be to strengthen relationships with current customers or to increase brand awareness. For still others, it may be to implement a completely new menu.

In most cases, a marketing plan is developed to cover a calendar year or, at least, a large portion of a year. Since the marketing needs of individual foodservice operations vary, no "one-size-fits-all" marketing plan can be implemented. Foodservice operators must develop, implement, and evaluate their own unique marketing plans.

A well-conceived and executed marketing plan can yield a tremendously positive impact on a foodservice operation and its profitability. By definition, a marketing plan indicates future actions. While no foodservice operator can predict the

Key Term

Marketing plan: A written plan that details the marketing efforts of a foodservice operation for a specific time (usually annually).

Key Term

Marketing strategies: The broad and long-term marketing goals a foodservice operation wants to achieve.

Key Term

Marketing tactics: The specific steps and actions undertaken to achieve marketing strategies (goals).

future perfectly, there are several advantages that can result from developing a formal marketing plan:

1) Improved assessment of the business environment

The creation of a formal marketing plan requires foodservice operators to carefully consider the needs of their target markets, the abilities of competitors, and numerous other factors that directly impact their business. These include an assessment of the social, economic, legal, and technological issues that directly impact how the businesses will be operated.

A careful examination of the overall business environment helps foodservice operators identify realistic and achievable marketing goals and strategies. A formal assessment of target market needs (that may regularly change!) also helps ensure that operators continually remain customer-focused.

2) Improved decision making

After foodservice operators carefully evaluate their business environments and the needs of their guests, they are in a better position to identify specific factors that directly affect their operations.

For example, if guests are requesting items not currently on the menu, changes may be in order. If new items are to be added to the menu, how will this information be communicated to current and potential guests? Foodservice operators can choose from a wide variety of alternatives as they prepare and send marketing messages. Note: in this example, some alternatives may be better suited for communicating the introduction of new menu items than others.

When marketing strategies have been carefully identified, it becomes easier to select the tactics that directly support those strategies. Given many alternative choices in marketing efforts, operators who create formal marketing plans can better decide which specific tactics will contribute most to the achievement of their marketing goals.

3) Better financial planning

Since formal marketing plans should identify the timing of specific marketing activities and their costs, foodservice operators who formalize marketing plans can better manage their budgets. For example, assume that a foodservice operator knows that the busiest time of the year is during the summer months, and their slowest time is the winter. Formal marketing plans can schedule high-cost advertising expenses in the summer months when revenues are strongest, and lower-cost marketing activities might be planned for the winter months.

4) Improved accountability

In addition to identifying marketing activities to be undertaken, a formal marketing plan identifies the specific individual or team responsible for implementing marketing activities. As a result, improvements in accountability for management or marketing staff performance are inherent in developing a formal marketing plan.

The best marketing plans indicate what should be done and when related activities are implemented. When specific marketing activities include the timeframe for completion, the accomplishment of planned activities on schedule helps to achieve an operation's marketing goals.

5) Improved assessment of marketing efforts

The strategic goals of a marketing plan should be realistic and measurable. When a strategic goal is realistic, a foodservice operation has a good chance of achieving it. Goals that are set unrealistically high can discourage operators because the goals simply cannot be attained regardless of their efforts. Strategic goals that are set too low, however, may cause operators to become complacent in their marketing efforts.

The strategic goals identified in a marketing plan must be measurable. Goals such as "increase business" or "improve customer satisfaction" can only be meaningful when they are measurable. For example, in a formal marketing plan a goal of "increase business" would better be stated as "Increase business by 15% on Friday nights." That goal is measurable, and an operator can determine whether marketing efforts helped to achieve that goal.

Similarly, a goal of "improve customer satisfaction" might better be stated as "improve the operation's average star rating from three stars to four stars on Yelp, Facebook, and TripAdvisor by December 31st of this year."

Again, that goal is measurable, and an operator can readily determine whether the goal was achieved during the time addressed in the marketing plan. The best marketing plans make it easy for foodservice operators to monitor their actual performance against their budgeted or expected performance.

A formal marketing plan helps foodservice operators better understand their guests, their competitors, and their overall business environment. It should be their goal to help develop a marketing mix (see Chapter 1) that guests will find to be highly attractive and enticing. Despite the challenges involved, the development of a formal marketing plan is essential to the success of all foodservice operations.

Creating the Marketing Plan

While the marketing plan for a foodservice operation will be unique to that operation, the basic steps utilized to create the plan are the same:

1) Determine marketing goals.
2) Identify target market(s).
3) Develop the marketing budget.
4) Select marketing channels.
5) Make task assignments.

These steps are previewed in Figure 3.4.

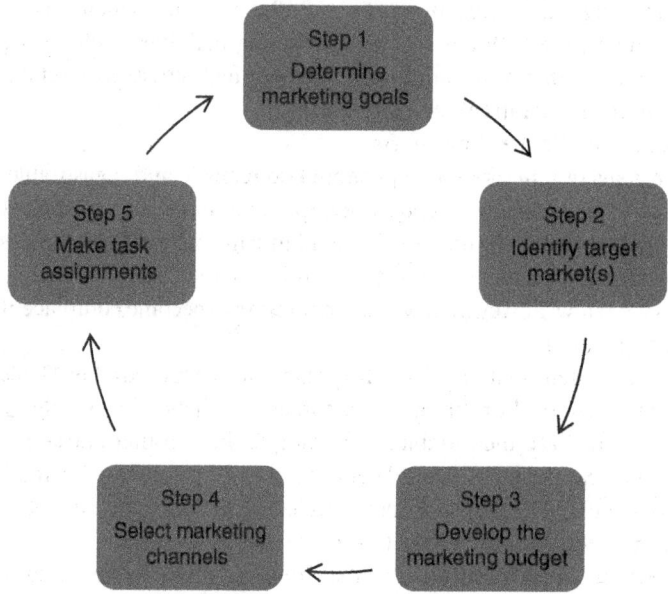

Figure 3.4 Steps in Marketing Plan Development

Step 1: Determine Marketing Goals

A foodservice operation's marketing goals should be achievable and measurable. In the first step of marketing plan development, foodservice operators may develop a written marketing plan that addresses a 6-month or 12-month period of time.

An important marketing goal for many operations is to increase revenue. If an increase in revenue is determined to be a major goal, operators must determine how much additional revenue they want to generate, the source of that revenue, and the time frame within which it will be achieved. For example, a casual full-service restaurant may have a goal to "increase beverage sales by 10% compared to prior year." If management determines this is a reasonable goal, its achievement (or failure to achieve it) can easily be measured.

The achievement of some marketing goals can be more difficult to assess. For example, a foodservice operator may wish to "improve the ease of the online ordering system." While the achievement of this goal is more subjective, it can, however, be measured. One method is to use a consumer survey that asks users of the on-line ordering systems to assess whether a newly implemented ordering process is easy to use.

Step 2: Identify Target Markets

In this step, operators carefully review the target markets they wish to serve. This is an especially crucial step for operators who identify more than one marketing goal. The reason: if this is the case, a part of the operation's marketing message may be directed to a specific target market, and another marketing message can be developed for another target market.

The identification of an operation's target market is also essential because, if substantial money and/or staff efforts focus on the wrong market, these financial resources will likely be misdirected and ineffective. Experienced foodservice operators know that their target markets can change. New target markets for foodservice operators may appear regularly, and their sequence of importance can also change.

Step 3: Develop the Marketing Budget

In its simplest terms, a marketing budget reflects the amount or percentage of sales an operation will devote to marketing efforts. For example, assume a foodservice operation forecasts sales of $1,000,000 during the time covered by the marketing plan. Also assume the operation will spend 5% of its revenue on marketing. The marketing budget for the operation would then be $50,000 ($1,000,00 forecasted revenue × 0.05 marketing budget = $50,000).

For those operations that have not yet opened, **preopening marketing expense** forecasts will be identified in the operation's initial **business plan**.

A new foodservice business may require significantly more marketing effort and expense before opening than would an existing and well-established operation such as another outlet in a chain property.

While many foodservice advisors suggest that the average operation should spend 3–6% of its projected revenue on marketing, special circumstances may dictate that as much as 10% (or more) of an operation's projected revenue might be devoted to marketing.

Step 4: Select Marketing Channels

Marketing channels can be broadly categorized as being either traditional or web-based. In this step of marketing plan development, operators must select specific marketing channels they believe are best able to deliver the marketing messages that will be created.

Key Term

Preopening marketing expense: A nonrecurring promotional or advertising cost incurred before a new operation opens. Also referred to as a "start-up" marketing cost.

Key Term

Business plan: A document that summarizes the operational and financial objectives of a business.

Key Term

Marketing channel: A method or form of delivering a marketing message to an operation's current and potential guests. Also commonly known as a "communication channel." Examples include print, broadcast, and web-based communication methods.

Current popularity, cost, and ease of ad development are all factors that will impact an operator's decision about the specific marketing channels to be selected for use.

Step 5: Make Task Assignments

In large foodservice operations and those with multiple units, one or more marketing specialists may be tasked with completing all the operation's marketing efforts. In smaller operations, marketing tasks may fall to an operation's owner, manager, or assistant manager.

Task assignment is a critical aspect of marketing plan development. It identifies who will complete the marketing task(s) to be undertaken and when the task should be completed. It is in this stage of market plan development that operators create their **marketing calendars**.

Marketing calendars should be specific, but they should also be flexible to allow an operator to change tactics as new circumstances dictate.

Key Term

Marketing calendar:
A schedule detailing all marketing activities planned for the time period covered by the marketing plan.

Technology at Work

Every foodservice operator who develops a marketing plan should include a formal marketing calendar in the plan. Today's foodservice operators have many tools available as they create their marketing calendars. Whether they choose to produce their calendars using Google, Excel, or Word documents, templates that can help them get started can generally be downloaded from the Internet at low cost or no cost.

To see some of the many options available to foodservice operators, enter "free marketing calendar templates" in your favorite browser and review the results.

What Would You Do? 3.1

"I think we're too late," said Mateo.

"But it's only the middle of October," replied Gabriela. "Christmas is more than two months away. How can we be too late?"

Mateo and Gabriela were discussing potential holiday party bookings that would utilize their operation's banquet room. The banquet room could hold up to 150 people and was a perfect location for employee holiday parties hosted by businesses. One of Mateo and Gabriela's marketing goals for the year was to increase their number of holiday party bookings.

"Well," replied Mateo, "I talked to our marketing consultant today, and she can help us produce a very nice print ad for mailing to local corporations in only a week or two."

"So," replied Gabriela, "That's good. We can be sending the ad out by early November. What's the problem?"

"The problem is that the holiday meeting planners I've talked with have already booked their space for this year," said Mateo. "With most holiday parties being held as early as the first weekend in December, I think we should have started contacting our target market for holiday parties back in July or August at the latest!"

Assume you were Mateo and Gabriela. How could a formal marketing plan that includes a calendar of marketing activities help to prevent the timing challenge they now face? What would you suggest they do for next year?

Implementing the Marketing Plan

As they finalize and implement their marketing plans, foodservice operators address several key issues. Among the most important are determining:

✓ What will be done?
✓ When it will be done?
✓ Who will do it?

Determining What Will Be Done

When making the final determinations about which marketing tactics will be included in their marketing plans, operators must ensure they stay within their marketing budgets and optimize the use of their marketing dollars.

In most cases, a marketing plan includes the use of multiple marketing channels. For example, an operation may implement marketing tactics that require the use of traditional marketing channels such as broadcast or print and a variety of web-based communication channels including websites, social media, and blogs. In many cases, an operation will also implement various on-site marketing activities such as the use of in-house promotions, revisions of existing menus, and training of service staff.

As they make final determinations about marketing channels to be used, foodservice operators must ask themselves several key questions about each potential channel:

1) Will it effectively reach my target audience?
2) Will it be cost-effective given the amount budgeted for its implementation?

3) Does it effectively communicate the marketing message?
4) Can its results be measured?
5) Is its use complementary to other marketing channels selected for use?

Determining When It Will Be Done

Each marketing strategy (goal) in a marketing plan will require that one or more specific tactics be completed to achieve the goal. Each chosen tactic should have a specific target completion date. This is important to monitor and ensure marketing activities are completed on time.

Figure 3.5 illustrates the use of a target completion date assigned to a specific marketing tactic that is chosen to support a specific strategic marketing goal. A completed marketing plan will include one three-column entry for each marketing goal established.

Determining Who Will Do It

In some cases, the completion of a marketing tactic will require the efforts of a single individual. This would be the case, for example, when an operation's owner chooses to work alone with a radio station's salesperson to produce advertising spots that will be broadcast during the time covered by the marketing plan.

In other cases, however, the completion of a marketing tactic may require team effort. For example, this could be a marketing tactic when introducing a new menu item that will be promoted for a specific time. The successful introduction of the new menu item will likely require the efforts of the operation's purchasing staff to ensure the item's ingredients are in inventory. As well, the kitchen's production staff should be involved as they are trained so the new menu item is prepared properly. The dining room's service staff must also be able to address guests' questions about the new item's preparation style and/or ingredients.

When the implementation of a selected marketing tactic involves the use of a team, it is best to designate a team leader who will be responsible for coordinating the efforts of those involved in successfully completing the tactic. In all cases, the

Strategy/Goal	Tactics	Target Completion Date
Increase, within three months, total revenue generated between 4:00 p.m. and 6:00 p.m. by 10% on a year-over-year comparison basis.	1) Produce a full-color table tent for in-house use, announcing discounted appetizers to be sold during the targeted time period.	June 1, 20xx; one month before this promotion's implementation so staff can be properly trained in its implementation.

Figure 3.5 Marketing Tactics Completion Schedule

implementation of a selected marketing tactic should be implemented or coordinated by a single individual who can then be held accountable for the tactic's implementation.

Evaluating Marketing Plan Results

Evaluation of an operator's marketing plan is a management process that examines the extent to which marketing strategies and goals identified in a marketing plan were achieved.

When marketing goals are measurable, a foodservice operator can calculate the **variance** between targeted financial results and actual results. Variances between revenue (income) and marketing costs (expenses) should also be calculated.

Key Term

Variance: The difference between an operation's actual and its estimated income or expense.

A revenue variance identifies the financial difference between the sales a foodservice operator estimated would be generated from a marketing strategy and the sales level the strategy actually generated. The formula used to calculate a revenue variance is:

Actual revenue − estimated revenue = revenue variance

For example, assume that a foodservice operator's marketing plan estimates that a specific marketing effort would generate an additional $50,000 in revenue within the next six months. At the end of the six months, the operator calculates that $40,000 in additional revenue was generated.

Using the revenue variance formula, this operator's variance would be calculated as:

Actual revenue − estimated revenue = revenue variance

or

$40,000 actual revenue − $50,000 estimated revenue = −$10,000 revenue variance

Key Term

Percentage variance: The percentage difference between an operation's actual and estimated income or expense.

When analyzing their marketing results some foodservice operators prefer to calculate a **percentage variance**.

The formula used to calculate a revenue percentage variance is:

$$\frac{\text{Actual revenue} - \text{Estimated revenue}}{\text{Estimated revenue}} = \text{Percentage variance}$$

As noted earlier, a percentage variance in revenue or expense is obtained by subtracting estimated revenue from actual revenue and then dividing the resulting number by estimated revenue. When calculating a percentage variance using numbers from the above example, the formula would be

$$\frac{\$40{,}000 \text{ actual revenue} - \$50{,}000 \text{ Estimated revenue}}{\$50{,}000 \text{ estimated revenue}} = \text{Percentage variance}$$

Or

$$\frac{-\$10{,}000 \text{ variance}}{\$50{,}000 \text{ estimated revenue}} = -0.20 \text{ (decimal form) or } -20\% \text{ (common form)}$$

In this example, the foodservice operation overestimated the revenue generated by 20%. Note that when using the variance percentage formula, the resulting decimal percentage can be converted to the more frequently used common form by moving the decimal point two places to the right or by multiplying it by 100. Also, observe that the calculation of a revenue percentage variance can result in either a negative or positive percentage.

For each measurable marketing tactic included in the marketing plan, foodservice operators should calculate the variance in either number of dollars or percentage variance of the actual results when compared to forecasted results.

When a marketing tactic produces greater results than forecasted, an operator may want to consider utilizing that tactic again. If, however, a tactic did not yield the expected results, the operator may want to reconsider (i.e. eliminate) the use of that tactic in the future.

With a well-thought-out and written marketing plan in place, foodservice operators can use their marketing budgets to increase their customer base and grow their revenues. However, marketing plans are not the only tools foodservice operators have available as they optimize their revenue and their success.

The most successful foodservice operators increasingly recognize the importance of the Internet to communicate with their guests. Most marketing experts agree that, today, the single most important marketing tool for all foodservice operations is their own proprietary website. For that reason, effective web-based marketing on an operation's proprietary website will be the sole topic of the next chapter.

What Would You Do? 3.2

"But we have placed an ad in the 'Welcome to Our Town' magazine for the last five years," said Mike Allen, manager of the Chicken Bucket, a quick service restaurant featuring fried chicken and traditional sides such as mashed potatoes and gravy, coleslaw, and biscuits.

Mike was talking to Denise Berger, the operation's assistant manager. Denise had proposed using the advertising budget previously spent for an ad in the local Visitors' Bureau magazine to update the restaurant's website. While Chicken Bucket's website was professionally done, it did not show any videos of the operation's products or customers. It contained only print messages and a few still photos.

"I think it's more important to keep our website up-to-date than to keep advertising in a print magazine. I know we should support the local Visitors' Bureau's efforts to promote our town, but I think it's more important to promote ourselves!" said Denise.

Assume you were Mike. How would you go about deciding whether to continue placing a print ad in a community-wide magazine versus investing the same amount of money in up-dating your website? How important would it be for Mike to carefully consider the implications of each alternative before making his decision?

Key Terms

Marketing	Marketing message	Business plan
4 Ps of Marketing	Marketing plan	Marketing channel
Marketing mix	Marketing	Marketing
Curb appeal	strategy	calendar
Market	Marketing tactic	Variance
Target market	Preopening	Percentage
Green practices	marketing expense	variance

Operator's 10-Point Tactics for Success Checklist

Evaluate your need for, and the current status of, each of the following operational tactics. For those tactics you think are important, but not yet in place, develop an action plan for its implementation including who will be responsible for the tactic's completion and the target date by which it should be completed.

				If Not Done	
Tactic	Don't Agree (Not Done)	Agree (Done)	Agree (Not Done)	Who Is Responsible?	Target Completion Date
1) Operator recognizes the importance of effective marketing to the successful operation of their business.	___	___	___		

(Continued)

Tactic	Don't Agree (Not Done)	Agree (Done)	Agree (Not Done)	If Not Done	
				Who Is Responsible?	Target Completion Date
2) Operator can state the importance of "Product" in the marketing mix of a foodservice business.	____	____	____		
3) Operator can state the importance of "Place" in the marketing mix of a foodservice business.	____	____	____		
4) Operator can state the importance of "Price" in the marketing mix of a foodservice business.	____	____	____		
5) Operator can state the importance of "Promotion" in the marketing mix of a foodservice business.	____	____	____		
6) Operator recognizes the need to identify the target market(s) that are most important to their business.	____	____	____		
7) Operator understands the process that is used to identify the needs of a target market.	____	____	____		
8) Operator can identify the information needed to create a comprehensive marketing plan.	____	____	____		
9) Operator recognizes the main components of a marketing plan will identify what is to be, when it will be done, and who will do it.	____	____	____		
10) Operator can state the importance of evaluating the results of a marketing plan after it has been implemented.	____	____	____		

4

Web-Based Marketing on Proprietary Sites

What You Will Learn

1) How to Choose a Primary Domain Name
2) How to Choose a Web Host
3) How to Create a Proprietary Website
4) The Importance of Search Engine Optimization
5) The Importance of a Social Media Presence

Operator's Brief

In this chapter, you will learn that in today's foodservice environment, an operation's proprietary website is its single most important marketing tool. The process of developing an effective proprietary website begins with identifying an operation's primary domain name, and this must be done very carefully.

In most cases, a foodservice operation will not buy its primary domain name outright. Instead, it can be rented for a fixed time. In many cases, the rental fee for a domain's name will be waived when an operator selects a web host: the organization that will maintain and deliver the content to be displayed on an operator's proprietary website.

Typically, a foodservice operator does not solely undertake the design and creation of an effective proprietary website. Instead, it will be created in conjunction with professionals who specialize in website development. Sometimes a foodservice operator's website development is undertaken in conjunction with an operation's web host, but this does not always occur. Regardless of who helps, operators must develop their proprietary websites with a focus on design, content, and the ability to track website traffic.

(Continued)

An effective website will include some mandatory and, in many cases, some optional features.

The optimization of search engine results is a critical aspect of website design. The selection of keywords and phrases will have a direct impact on an operation's placement on a search engine results page (SERP) so these terms must be chosen carefully.

In addition to the content placed on their proprietary websites, operators increasingly place information on social media sites. The popularity of such sites continues to grow, and the utilization of these sites should be an important part of every operation's overall web-based marketing efforts.

CHAPTER OUTLINE

Choosing the Primary Domain Name

To establish a web presence, every foodservice operation must begin by selecting its **domain name**.

Foodservice operators need not be computer experts to understand the importance of their domain names. Essentially, computers rely on a language when they communicate with each other, and each computer device is identified with its own unique **IP address**. For example, a foodservice operation's actual IP address might look like the following:

192.168.123.132.
or
3ffe:1900:4545:3:200:f8ff:fe21:67cf

Key Term

Domain name: The unique name that appears after "www" in web addresses and after the "@" sign when e-mail addresses are used.

Key Term

IP address: A unique address that identifies a device on the Internet or a local network. IP stands for "Internet Protocol," a set of rules governing the format of data sent via the Internet.

For most people, typing in a long IP address to connect to a website is not realistic. Actual IP addresses are complex and not easily remembered, and that is why domain names were created: to convert IP addresses into something more easily remembered. Therefore, an operation's domain name can be considered a "nickname" for its IP address. Put another way, domain names provide a human-readable "address" for foodservice operations, and the address chosen must be unique: it cannot be an address already in use by another person or entity.

Selecting a domain name to help market a foodservice operation must be done thoughtfully, and there are several key steps and tips. These include:

1) Choosing an appropriate **extension**

Key Term

A foodservice operation's chosen domain name will be followed by an extension (examples include those followed by .com, .net, and .org.)

Extension (domain name): The combination of characters following the "period" in a web address.

Today, there are hundreds of domain name extensions available in addition to ".com" which is the original and most common extension. Despite the facts that both ".rest," and ".restaurant" exist and are used by some operations, in most cases, it is still best if an operation uses ".com" as its extension. The reason: most guests searching for a restaurant are familiar with ".com" and may not be familiar with other extensions. In fact, many users, especially those who are not very tech-savvy, automatically type ".com" at the end of all domains without even thinking about it.

2) Keeping the domain name short

Some foodservice operators want to put as many words and as much information in their domain name as possible. For example, a New York-style pizza operation named "Tony's" might want to use a domain name such as:

"Tony's, home of the best New York-style pizza in the city.com"

This domain name is descriptive and includes the operation's name and a statement about the quality of the pizza sold by the operation. However, the name is too long and would not be easily remembered by most potential guests. Note also that entering the operation's web address requires the use of an apostrophe between the "y" and "s" in the name "Tony's," a comma after the word "Tony's," and a dash (hyphen) between New York and style. Many potential users would not know about these requirements to find the website's address.

Also, Internet marketing experts typically recommend that a domain name consist of 15 or fewer characters. The reason: when a domain name is too long, users increasingly make mistakes (typos) when entering the longer name.

In the above example, a better domain name choice might be:

Tonyspizza.com

If that domain name is already taken by another operator, how about:

Trytonyspizza.com

Note that both domains are short, simple, and would likely be remembered by most guests.

3) Keeping the domain's name easy to pronounce and spell

When utilizing the name of a foodservice operation in a web address, the domain name must be easy to pronounce and spell. Operators selecting their domain names should avoid foreign words and language terms unfamiliar to many Internet users. Another key suggestion is to avoid all "hard-to-spell" words.

4) Avoiding hyphens and double letters

Some operators may need to create their own unique domain main name when their first choice has been selected by another operation. For example, if a donut shop operator named Tim finds that the domain "Timsdonuts.com" is already in use, the operator may be tempted to select a domain name of "Tims-donuts.com."

This domain name would likely be available, but it is likely that most guests looking for "Tim's Donuts" would go to the original domain name site, rather than the operator's hyphenated site (hyphenated domains are also prone to typos). Similarly, domain names with double letters are prone to user entry typos. In most cases, the avoidance of double letters and symbols makes a domain name easy to remember, easy to type and, therefore, more effective.

5) Researching the proposed name

Before finalizing a domain name, operators must determine if there is already a business using the same name. If so, the operation will not be allowed to use the name. When researching the availability of a domain's name, operators should perform a Google search and check for the name's availability on popular **social media** websites.

Key Term

Social media: A collective term for websites and applications that enables users to create and share information and content or to participate in social networking.

Technology at Work

In most cases, the use of high-quality domain names is not free. A domain name must be rented or, in some cases, purchased. Typical rental fees for a domain name range from $2 to $25 per month.

There are many organizations that assist operators in searching the Internet to determine whether a proposed domain name is already used, or if it is available for use. They can also provide suggestions about alternative domain names that are memorable and available.

To see how these search sites work and to learn how to use them, enter "Search for available domain names" in your favorite search engine and view the results.

After an operator has selected an available domain name, the next step in using it is its **registration**.

Domain name registration is required for a website, e-mail, or other web services. An operation's domain name can be used as long as the rental payments continue for it. In most cases, a **web host** will waive rental fees for a domain's name if an operator agrees to select that organization to host its website.

Key Term

Registration (domain name): The act of reserving a domain's name on the Internet for a fixed time (usually one year). Registration is necessary because domain names cannot typically be purchased permanently.

Technology at Work

Domain name registration gives a foodservice operation a recognized identity, and many organizations can register an operation's chosen domain name. After a domain name is registered, the information about its owner is publicly available.

Domain names are registered by registrars accredited with the Internet Corporation for Assigned Names and Numbers (ICANN): the nonprofit organization responsible for coordinating all numerical spaces on the Internet.

To see how these registration organizations work and to learn about the services they offer in addition to domain name registration, enter "How to register domain names" in your favorite search engine and view the results.

Choosing the Website Host

The web host selected by a foodservice operator will become an important marketing partner, and Figure 4.1 illustrates why this is true. A web host does more than merely receive content from a foodservice operator. It also distributes that information to those visiting the operator's website. The ability to effectively manage both key tasks is the mark of a high-quality web hosting partner.

Key Term

Web host: An organization that provides space for holding and viewing the contents of a website. All websites and e-mail systems must be hosted to connect an operator's proprietary website content to the rest of the Internet.

Web hosts play a key role in disseminating information about a foodservice operation. First, they store information about an operation in web servers that are connected to networks. When a visitor types in a web address, the Internet transmits the request to the web host. The host then retrieves the operation's files and sends the information back to the visitor's computer so the site can be viewed.

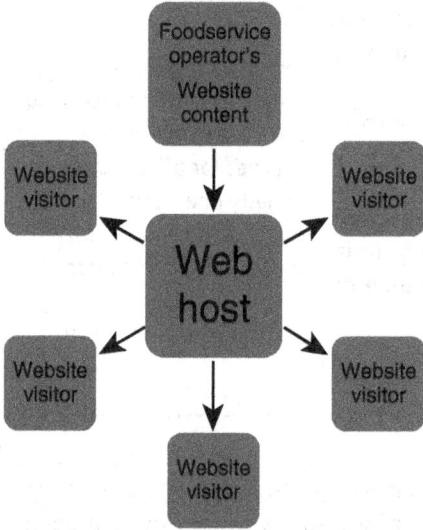

Figure 4.1 Web Content Distribution

There are important factors to be considered when selecting a web host:

1) Speed

The speed at which an operation's website loads is important for making it easy to use. Today's foodservice guests will have little patience with a slow-loading website. Speed of download is especially important if a foodservice operation includes videos on the site because these files take longer to download than information presented in regular or **PDF** text.

Key Term

PDF: Short for "Portable Document Format:" a document format that allows this information to be shown clearly regardless of the software, hardware, or operating system used by the viewer.

2) Reliability

The typical foodservice operator will not have detailed knowledge about the specific servers and networks used by a web host. However, the chosen web host should be reputable and able to guarantee consistently high levels of speed, reliability, and minimal downtime.

3) Security

The security levels offered by a web host are extremely important, especially when a website will process guest payments or send personal information to the website. (Consider, e.g. when guests participating in a guest loyalty program enter personal data on the website to join and/or redeem their rewards.)

4) Support

The operation's website is an increasingly important marketing tool, and problems such as an inability to download content, blank or missing pages, and related issues must be resolved quickly. The best web hosts have 24/7 support that is easily accessible and that can quickly lead to the resolution of content display issues. As an operation's website increases in content size and functionality, the importance of high-quality support increases as well. The best web hosts offer a choice of support contact methods including live chat, telephone, and e-mail.

5) Data Management Tools

The best web hosts also provide operators with data management tools some of which are at no cost (they are included as part of the web host's monthly fee) and in other cases they are available for purchase.

Essential tools that should be provided by a web host include:

✓ A Content Management System (CMS) program used to easily make changes in website content
✓ A Customer Relationship Manager (CRM) program to manage customer loyalty programs
✓ A website visitors' tracking system to gain important information about a site's users
✓ E-mail services (optional), but which are often included at no charge

Many web hosts offer foodservice operators a package that includes hosting and domain name registration. Some operators purchase their hosting and domain name registration from the same company. One possible concern in this approach can occur if the operator wants to use another web host in the future. Then it can be difficult to separate the domain's name from the original web host. Some Internet-marketing experts recommend securing the domain name separately so web hosts could be easily changed if an operator desires to do so.

Technology at Work

GoDaddy, Amazon Web Services, and Google Cloud* are currently the largest web hosts. These may fit the needs of a specific foodservice operator and should be considered as potential web hosts. However, there are likely other web hosts that might better address the operator's needs.

The best web-hosting decisions are made when operators know what their website should be able to do. For example, will guests visiting the website be able to do online ordering and payment, or will they simply be viewing a menu in a print (PDF) format? Similarly, will guests be able to make a reservation on the website or join and review guest loyalty program points?

The features and functions to be offered on a foodservice operation's website most often help determine the best web host. To see a list of available foodservice-oriented web hosts that could be considered before making a final selection, enter "Best web hosts for restaurants" in your favorite browser and view the results.

Ranking as of 1-17-2024: https://www.hostingadvice.com/how-to/largest-web-hosting-companies/

The Proprietary Website

Key Term

Proprietary website: A website in which the foodservice operator controls all the website's content and can readily make changes to it.

A foodservice operator's **proprietary website** is one in which the operator has full control over the website's content.

Regardless of their service style or size, all foodservice operators should create and manage a professionally developed proprietary website that their guests will find at:

www.registered operation domain name.chosen ext

To illustrate, the proprietary website address for Tim's pizza could be:

www.timspizza.com

The marketing-related goals of a foodservice operator's proprietary website are easy to identify, and these include:

1) Showing the operation's NAP

Key Term

NAP: An abbreviation for an operation's name, address, and phone number.

One function of a website is to show an operation's name, address, and phone number (**NAP**). This is important whether an operation is independent or part of a chain. If it is an independent and especially a start-up, potential guests must know the name of the business, where it is located, and how to contact it. If a foodservice operation is part of a local or national chain, a proprietary website identifies the specific operation's location. If the operation is a food truck with regular location changes, information about future locations and dates would be included here.

In years past, foodservice operations disseminated their NAP information in the Yellow Pages, a printed directory distributed by telephone companies. Today's customers search for NAP information on their computers, tablets, and smartphones. A proprietary website enables them to learn about a business and what it offers anytime and at any place they choose to do so.

2) Attracting customers

Successful foodservice operations maintain their current customer bases and expand them with new customers. They know that reaching thousands of potential new customers using only traditional marketing methods such as direct mail, print advertising, and/or radio or TV can be time-consuming and expensive.

A strong online web presence enables operators to reach more people locally while paying a relatively modest amount for exposure. In most cases, potential guests prefer to look online rather than call a business to learn basic information such as operating hours or location. Operations without active websites miss out on potential customers who do not want to make a phone call to ask about offered products and services.

3) Building credibility

Many customers using the Internet to research a business have likely viewed thousands of different web pages from hundreds of businesses, and they can easily differentiate between a high-quality website and one that is poorly designed. The quality of an operation's website communicates much information to those surfing the web.

When a website appears professionally designed and is easy to navigate and visually exciting, it enhances a foodservice operation's image. Guests viewing the website often believe a professionally prepared website suggests the operation is run professionally and is established and successful. This straightforward communication of business stature builds an operation's reputation and enhances its credibility.

4) Growing the business

In many cases, a potential customer will not know the exact domain name of an operation they desire to visit, and the operation's proprietary website then becomes particularly important. To illustrate, assume a family visiting an unfamiliar city wants to visit an Asian-style restaurant that offers Thai cuisine. By performing a **Google search,** this family would be directed to one or more proprietary websites created by various foodservice operations in the city that offers Thai cuisine.

Key Term

Google search: Also known simply as "Google," this is a method of finding information on the Internet. Note: Google conducts over 92% of all Internet searches and is the most visited website in the world.

After reviewing the information on alternative websites, the family decides which restaurant to visit. Their decision may be based on price or location or, in many cases, on the professional appearance of the website itself. Conversely, if a Thai foodservice operator in the city did not have a proprietary website, the travelling family would not know about this operator nor be able to select it from available alternatives.

5) Gaining a competitive advantage

Some foodservice operators believe that, if they have a Facebook page or are otherwise utilizing social media, a proprietary website is not important, and this is an error. Note: The importance of a social media presence is addressed later in this chapter. However, it is essential to understand that a foodservice operation's ability to communicate to their guests on social media platforms is typically restricted in terms of design, process, and technology.

A foodservice operator desiring to provide specific information and services to guests typically recognizes that a proprietary website assists them to do so while gaining a competitive edge. The reason: a website not only attracts new guests, but it can also be used to better serve existing guests. The ability to update operating information such as new menu items and specials on a 24/7 basis improves current guest communication, helps build customer loyalty, and enhances an operation's professional image.

There are several ways to examine key components of an operation's proprietary website, and one useful method is to assess three important website characteristics:

1) Website Design
2) Website Content
3) Website Traffic Tracking

Website Design

Few foodservice operators have the expertise and time required to design their own proprietary website. In some cases, website design assistance is offered by a web host offering basic website templates. Alternatively, operators can utilize the services of a professional website designer or design organization.

Professional web designers implement several tactics to design dynamic and effective websites, and they pay particular attention to three key areas:

✓ Appearance
✓ Reading pattern
✓ Navigation

Appearance

The overall appearance of a proprietary website is determined primarily by color, type-style, and imagery. The basic colors used in website development should fit with an operation's service style and, in most cases, should be limited to five colors or less. The colors used on the website should be complementary. If colors are used to identify an operation's **brand** either in its uniforms, logo, and/or signage, these are normally the website colors that will best reinforce the operation's brand image.

Key Term

Brand: The specific ways in which a foodservice operation differentiates itself from its competitors.

The type-style used to create the website should be of a size that is easily read and especially so when an operation's menu is posted online.

The imagery used in a professional website can vary from high-resolution photographs to full videos. The imagery used in all forms of graphics should be expressive and capture the personality of the company. Since a website visitor's initial impressions are most often formed visually (not by reading printed content), the imagery used on the website is of extreme importance.

Reading Pattern

The most common way for visitors to scan text on a website is by using an **F-shaped scanning pattern**.

Figure 4.2 is a visual representation of the F-shaped scanning pattern utilized when viewing information displayed on a smartphone.

Professional website designers are very familiar with the F-shaped scanning pattern and present website content in a way that is complementary to the reading style of the website's users. It is especially important to consider the F-shaped scanning pattern when designing websites to be mobile-device friendly since these are the types of devices increasingly used to view websites.

Navigation

Navigation refers to the process of users' finding information of importance to them when they visit a website. When websites are easy to navigate, the **user experience** is positive. Then viewers can readily find important information or accomplish a desired task (such as placing an order or making a reservation) with relative ease.

Proper website navigation contributes to a positive user experience by presenting a website layout that is intuitive to use and which allows users to find the information they want as quickly as possible. When navigation on a website is poor, users will find the website confusing, often give up their search for information and look elsewhere for what they need.

Key Term

F-shaped scanning pattern: A text scanning pattern characterized by fixations concentrated at the top and the left side of a page. Website viewers typically first read in a horizontal movement across the upper part of the content area. Notes: These areas initially form the F's top bar, and viewers then scan down. The "F" in F-shaped scanning pattern is short for "Fast," and the method is also commonly referred to as the "F-based reading pattern."

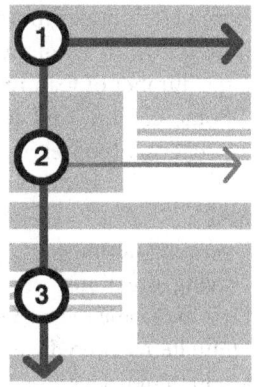

Figure 4.2 F-shaped Scanning Pattern

Key Term

Navigation (website): The process of moving from one content section of a website to another section of the same website.

Key Term

User experience: The feelings resulting from a user's interactions with a digital product.

Most professional web designers working with restaurants recommend that menus be split into lunch and dinner if they differ. In addition, separate wine beer and spirit menus to keep the food menus from seeming to appear too overwhelming and "too busy." Design features that include a "call-to-action" button that encourages guests to book a table, order online, or buy a gift card, must also be easy to use if they are included on the website.

If an operation has multiple locations, the proprietary website should include links specific to the location for which the customer is searching. Different directions and phone numbers, menus (if they vary), and online ordering features must be clearly identified. Then, for example, guests do not place a pickup order with "Tony's Pizza South," when they intended to place their order with "Tony's Pizza North."

Website Content

The actual content on a proprietary website can vary widely based on the needs and interests of the operation and its ownership. In some cases, a website offers extremely limited information consisting only of the operation's NAP, and perhaps a PDF of its menu(s). In other cases, a proprietary website may include a large number of features and functions.

Some content and features might be considered mandatory because guests expect every foodservice operation's website to address them, and these include:

NAP
Operating hours
Driving directions
Link to Google maps (to show an operation's map location)
Menu items offered
Menu item prices
Payment forms accepted
Links to social media sites (if any)

Other content areas may be included on the website if an operator determines them to be of value. Examples of optional content areas can include:

Online Ordering
Online ordering is not yet considered a mandatory content area for all operations. However, many foodservice operators believe this feature is or soon will be required on the foodservice's website (or an app). Designated payment apps are becoming more necessary because guests prefer using apps on mobile devices.

The reason: it is easier for them to just tap on an icon located on their mobile screens instead of looking for a website with web browsers.

While the future of takeout and delivery ordering by guests is still being written, operators learned during the COVID-19 pandemic that they absolutely need a way for guests to place orders without being on-site as they do so.

The superiority of online ordering versus telephone-only ordering is continually proven. One reason is because an online digital record of the order reduces guest misunderstandings, confusion, kitchen errors, and payment disputes. An online ordering function coupled with online payments can help manage malicious or mischievous order placers with no intention of picking up orders or who are placing unwanted orders for others.

Find Out More

Understanding customers is a big step toward selling more and growing a business. Successful businesses pay attention to how their customers behave and how they prefer to purchase products. Due to changing technology, foodservice customers continue to move toward mobile-based purchase options because they are convenient. As a result, increasing numbers of foodservice operators are creating their own operation (or chain) specific apps. A robust app that helps guests order food quickly and efficiently is increasingly vital. So too is a good loyalty program with the ability to book tables and offering multiple payment options.

Large segments of customers prefer using mobile apps for several reasons including:

1) *Ease of access:* As stated earlier, customers prefer using apps on mobile devices because it is easier for them to tap on an icon on their mobile screens than to look for a website through web browsers.
2) *Better user experience on mobile devices:* Some websites run very inefficiently on mobile phones because their complicated features do not fit well on mobile screens. This calls for creating a mobile platform that uses features of mobile devices to offer an improved user experience.
3) *Better engagement:* App features are generally designed to engage customers for a longer period of time.
4) *Ease of payment:* Mobile apps can include versatile and easy payment options like credit/debit card payment, mobile wallet integration, net banking, one-touch payment, and cash on delivery.

As more customers use their mobile devices and apps to do their business, foodservice operators must address those preferences. To remain updated about the changing ways foodservice operations can use apps to grow their customer bases, enter "App development for restaurants" in your favorite browser and view the results.

Special Order Pickup Instructions

An increasing number of guests order menu items for pickup in many foodservice operations. An operation accepting pickup orders can make it easy for guests to get them by providing special instructions about where to park, where to go, and who to see when takeaway orders are picked up. This feature is especially important during busy service times when arriving guests may not know special arrangements have been made for order pickups.

Online Table Reservations

This feature allows website or app users to make a reservation including date, time, and number in the party. When reservations are easy to make, site visitors have no need to go to another website. Note: The best of these reservation systems may even allow guests to choose their preferred table in the dining room.

Gift Card Purchasing

Gift cards are extremely popular both for the person purchasing them for their own use and for gift-giving. When an operation offers gift card purchasing online, they make it easy for a visitor to buy something they will likely need and want now or in the future. Note: When promotional discounts are offered (e.g. a $50 gift card can be purchased for $45), online gift card sales can be significant.

Customer Loyalty Program Enrollment

Operators with a customer loyalty program can offer website viewers the opportunity to enroll in the program online. Providing contact information such as an e-mail address and/or telephone number increases the size of the loyalty program's membership. It is also generally easier for the operator to send future communications to its most loyal guests.

Operators selecting this option should recognize concerns about data privacy issues and fear of fraud. For example, some persons may receiving text messages to be somewhat invasive. Note: While younger generations are often more accepting of receiving text communication, many in older generations are not.

Text messages are a direct way to ensure a customer sees an operator's messages. However, some guests are more likely to "opt out" of receiving these messages in the future since unsubscribing to text messages consists of just responding "No" to future messages. This differs from unsubscribing to an e-mail that is a longer (several step) process that many guests will not consider to be useful, time-effective tasks.

On- and Off-site Catering Information

For operations offering on-site banquets for large groups or who have mobile operations providing off-site catering services, information about these services should be posted online. Visitors to a foodservice operation's proprietary website should never have to wonder if they can make a large pickup order and/or host a large event at the operation.

An Online Promotion Feature

Online promotions can be offered only to guests who view the operation's website. These promotions can make the website's viewer feel "extra special," and promotions can be changed regularly to encourage visitors to return regularly to the website. This, in turn, further establishes the brand's online identity and enhances customer loyalty.

On-site Video

The adage that "A picture is worth a1000 words" applies to foodservice websites, and high-resolution photos of menu items and interiors are generally essential. Short, on-site videos of real guests enjoying real food make an especially powerful advertising statement. It also projects a level of professionalism in the foodservice operation that is not projected by photos alone. If used, these videos need to be short as they can be easily loaded for viewing on smart devices.

Guest Reviews

This optional feature is listed last, but not because it is last in importance! In fact, as most operators agree, guest reviews are an extremely important factor when considered by potential customers as they select a foodservice operation for the first time.

The importance of encouraging and displaying positive, user-generated reviews on a website and social media and the need to professionally respond to any negative reviews (see Chapter 5) are essential to an operation's success.

What Would You Do? 4.1

"I think we have to do it," said Jasmine. "If we don't go for it, we're going to continue to lose out." Jasmine, the manager of the Ham From Heaven restaurant, was talking to Tarique Fletcher, the restaurant's owner.

Ham from Heaven was famous for its slow-roasted ham sandwiches and side dishes. Business was good, and guests continued to place their orders online for pickup or delivery. Many of the delivery orders were placed on third-party delivery sites and apps such as Door Dash and Grubhub. In most cases, these orders were paid for at the time the order was placed, and guests really liked the convenience of being able to do so.

Customers who called in and then picked up their orders at the restaurant paid for them at the time of pickup because the Ham From Heaven website did not provide an online payment option when orders were placed directly with the restaurant. When the operation was busy, wait times for order pickups

(Continued)

were extended because, in part, it took time to process the individual payments for the pickup orders.

Jasmine recommended to Mr. Fletcher that they add an online payment feature to their operation's proprietary website. This feature would allow pickup order guests to pay for their items at the time they placed their online orders.

"How does that work?" asked Mr. Fletcher. "I mean, how is the sale automatically recorded in our POS system? And what about payment card security for our guests? How do we ensure that?"

Assume you were Jasmine. Do you think the questions asked by Mr. Fletcher are valid? What additional issues should Jasmine address as they consider adding this new feature (guest prepayment at time of order) to the operation's website?

Website Traffic Tracking

Traffic tracking tools can tell operators much about their proprietary website and the guests who have visited it. Traffic tracking is also important because it provides valuable information to modify and improve an operation's proprietary website.

Every operator with a proprietary website can benefit from tracking the site's visitor data. Today, Google Analytics is the best tool for website traffic monitoring. Many of its services are free and easy to use. Three important metrics that Google Analytics provides are:

✓ Average time a user spends on a website
✓ Sources that direct users to the website
✓ Interests of the website's viewers

Web traffic is carefully measured to help evaluate websites and the individual pages or sections within them. While several website metrics are important to operators, most marketing professionals suggest that, in addition to other metrics, the **click-through rates (CTR)** and the **bounce rates** of a website should be regularly analyzed.

Key Term

Traffic tracking (website): Information that tells website operators, among other things, the number of site visitors their websites have had, how the site was found, how long visitors browsed different pages on the website, and how frequently visitors were converted into customers.

Key Term

Click-through rate (CTR): The number of clicks an ad or promotion receives divided by the number of times the ad is viewed. The formula used to calculate CTR is:

$$\frac{\text{Number of Clicks}}{\text{Number of Views}} = CTR$$

For example, if an operator's promotional ad on a website had 5 clicks with 100 views, the CTR would be 5/100 = 5%.

Among additional website traffic metrics that are important and may be monitored are:

✓ The total number of website visitors
✓ The average amount of time each visitor stays on site
✓ The average length of time each page is viewed
✓ The most popular website pages
✓ The least viewed website pages
✓ The source of website visitors
✓ The geographic locations of visitors that provided the most online orders

Key Term

Bounce rate: The percentage of visitors who viewed only one page before exiting a website.

Search Engine Optimization

Every foodservice operator wants to develop a website that is vibrant, easy-to-use, and effective. To be effective, it must first be seen and, to be seen, it must be easily found on the Internet. In some cases, foodservice operators either do not understand or discount the importance of search engine results when their websites are created.

A **search engine results page (SERP)** is the list of results a search engine returns in response to a user's specific word or phrase query.

It is extremely important that an operation is near the top of a SERP when it is viewed. Fortunately, proper website design includes paying careful attention to **search engine optimization (SEO)** strategies and principles.

Marketing professionals with experience in SEO strategy can assist foodservice operators by ensuring their websites' placement on SERPs is as good as it can possibly be. This is critically important because, no matter how professionally it has been developed, a website can only serve as a positive marketing tool when users see it!

Key Term

Search engine results page (SERP): The webpage of a search engine such as Google, Bing, or Yahoo that shows a user when a user types in a search query.

Key Term

Search engine optimization (SEO): The process of improving and modifying an operation's website with the purpose of increasing its visibility when users search for products and/or services related to the operation.

How Search Engines Work

Foodservice operators need not be experts in SEO to recognize its importance. However, a basic understanding of how search engines return their results in response to a user query is important for the successful development of a high-quality website.

Essentially, search engines such as Google are continuously scanning webpage content placed on the Internet. The engines move from site to site, collect information about those pages, and develop it in an index. The index can be visualized as a giant library in which the librarian (the search engine) can immediately pull up a book (a webpage) that includes exactly the information requested by a library visitor (the computer user).

Next, algorithms analyze pages in the index by considering hundreds of signals and pieces of information provided on web pages. The reason: this is done to determine the proper order pages should appear in the SERP for a given computer user's query.

Returning to our library analogy, it is as if the librarian (the search engine) has read and can recall information from every single book in the library and can immediately tell a user which ones have the proper answers to their questions. Unlike advertisements, search engines will not accept money to have an operation's web page appear higher on a SERP than do other websites. SERPs can, however, be directly influenced by the **keywords** and phrases operators place on their proprietary webpages.

Every SERP produced is unique, even for search queries performed on the same search engine using the same keywords or phrases in their search queries. This is because search engines customize the experience for their users by presenting their SERP's results based on additional factors beyond the user's search terms. These additional factors can include the user's physical location, browsing history, and social media settings. For example, a user located in Biloxi, Mississippi, and entering the keywords "pizza restaurants" will generate a SERP that is very different from that of a user located in New York City who enters the exact same keywords.

Key Term

Keywords: The words and phrases that users type into search engines to find information about a particular topic. Also known as "SEO keywords," "Key phrases," and "Search queries."

Choosing Keywords and Phrases

Keywords are any term used in a search, and they can be a single word or a long phrase. For example, the word "Pizza" is a keyword, but so is the multi-word phrase, "Pizza near me now." As shown in Figure 4.3, anything typed or spoken into a search engine's query bar can be considered a keyword.

A detailed examination of how professional website designers collaborate with foodservice operators to optimize SERPs for their unique proprietary websites is beyond the scope of this chapter. There are, however, some principles that all operators can use as they place keywords and phrases into the content displayed on proprietary websites.

Figure 4.3 Google Keyword Search

To select keywords and phrases that assist in improved SERP placement, food-service operators should:

1) Think like customers.

Search engines are designed to be user friendly, and key words and phrases anticipate normal language and queries. When determining the best keywords and phrases, operators should ask themselves; "If I wanted to find my foodservice operation what would I type into Google?" Talking to other foodservice professionals, staff, and current customers can help generate opinions about words or phrases that would likely be used to search for a specific operation.

2) Consider search volume.

In some cases, it may be possible for a foodservice operation to be listed at the very top of a SERP, but it is not a SERP that many customers want to see. In this case, ranking first for a search word practically no one ever uses is an ineffective use of a keyword. Operators should avoid the extensive use of unfamiliar terms, and those that are hard to spell.

3) Use available keyword research tools.

There are a number of free SEO tools that can help operators gain insight into their website traffic. By using Google Analytics, Keyword Planner, and Google Search Console, for example, operators can gather data on keyword volume and current search trends. The use of keyword research tools can provide valuable information on how to choose keywords.

4) Study competitors' websites.

Nearly all foodservice operations have direct competitors. It is often worthwhile to look at these competitors' proprietary websites to see the keywords and phrases they are using. Doing so might help an operator to include terms or descriptions they might have otherwise forgotten.

5) Select keywords that are also suitable for social media use.

When an operation has a presence on social media (and most should!) using the same keywords on the proprietary website as on the operation's Facebook, YouTube, Instagram, WhatsApp, or Twitter (X) accounts can help to optimize search engine results on these sites.

> **Find Out More**
>
> Search engine optimization (SEO) and the selection of the best keywords to optimize a foodservice operation's placement on a SERP are both an art and a science, and SEO strategies are ever-evolving.
>
> SERPs are the web pages displayed to users when they search for something online using a search engine such as Google. The user enters a search query (often including specific terms and phrases referred to as "keywords"). The process of creating SERPs constantly changes because ongoing experiments conducted by Google, Microsoft, Bing, Yahoo, and other search engine providers can offer users a more intuitive and responsive experience.
>
> As search engine technology changes so will the ways SERPS are generated. To follow changing SEO technology and to stay updated about this increasingly important marketing area, enter "optimizing search engine results for foodservice operations" in your favorite browser and view the results.

Social Media Sites

The rise in popularity of social media sites makes them essential tools for successfully marketing foodservice operations. Just as operators can choose the content placed on their proprietary websites, they can also select some of that content for use in their online social media accounts.

The use of social media sites for social networking continues to grow. Increasingly, they are the sources where consumers can seek information about foodservice operations they may want to visit.

The increase in popularity of social media sites as a means to promote a foodservice operation is due to several important characteristics:

1) The sites are user-friendly: It takes very little technological knowledge to post content on a social media site. Therefore, most of the operator's customers can easily participate in social media regardless of the demographics of their target markets.

2) The sites provide communication opportunities with real people: The primary purpose of social media accounts is communication and interaction, and a social media account differs from a website. For example, a traditional website provides one-way communication, and social media users can be involved in two-way communication.

3) The sites allow the creation of groups: The most popular social media sites allow users to create groups with like-minded people who share similar interests, hobbies, and/or experiences. These groups act like clubs whose membership is open to anyone who shares these common traits. (Like most

clubs, those who participate most often receive the most value from the membership.)

4) The sites are free to use: Social media sites do not charge their users; rather they generate income through the sale of ads, and this maximizes their accessibility.

5) The sites post reviews of businesses. In nearly all cases, customer reviews of a business posted online are given more weight by readers than traditional advertisements placed by the same business. User-generated reviews of a business can increase or decrease the credibility of the business's products or services in the minds of potential customers before they make a purchase decision.

Over 300 million people in the United States use social media, and the typical user visits social media sites for two hours and 35 minutes every day.[1] Foodservice operators who wish to establish an online social media presence to reach these users should select one or more social media sites. They should spend their time focusing on quality posts, conversations, and ways in which they can foster engagement with potential customers.

Foodservice operators must create their own social media accounts before they can post desired content on a social media site. Doing so is usually easy and, in most cases, entails no costs other than the time involved in creating and maintaining the account. The popularity of any specific social media site may increase or decrease over time. Therefore, foodservice operators should monitor their own and competitive sites to stay abreast of those that are the most popular sites. They should also ensure that any content placed on their social media sites is accurate, up-to-date, and consistent with the information posted on their proprietary websites.

This chapter has addressed how foodservice operators can increase their business and their online presence using proprietary websites and the content they post on social media sites when they join. There are, however, other sites about which readers will find foodservice-related information posted by other businesses and an operation's business partners. Increasingly, this data about a foodservice operation is posted on **user-generated content (UGC) sites**.

In most cases, an operator does not have direct control over the information posted on third-party operated websites or on UGC sites, and they cannot readily change this information.

Key Term

User-generated content (UGC) site: A website in which content including images, videos, text, and audio have been posted online by the site's visitors.

However, online content about an operation posted on third-party managed sites and on UGC sites is increasingly viewed by potential guests. The successful

---

1 https://www.demandsage.com/social-media-users/#:~:text=Social%20media%20is%20used%20by,spot%20with%20302.25%20million%20users retrieved July 30, 2023.

management of the content appearing on sites of these types are the topic of the next chapter.

What Would You Do? 4.2

"I think you have your priorities all screwed up," said Otis.

"What do you mean by that?" asked Maggie.

Otis was Maggie's grandfather. He had been the owner and operator of Otis Candies, a small retail candy shop that specialized in handcrafted artesian chocolates. Otis had been operating the shop for over 40 years, and Maggie was spending her summer off from college helping Otis in the store. Thirty days after her arrival, online orders processed on the store's website had increased by nearly 50%.

"What I mean is we should be spending our time speeding-up the processing of all our online orders. That's where our business is increasing the most. I think you're wasting your time with all of these social media accounts that you started," said Otis.

Otis was right that Maggie had established an online presence for Otis Candies with three of the most popular social media sites. She enjoyed posting information and short videos about the shop's long history and how it makes its popular gourmet chocolate items. The sites had attracted a significant number of followers because many were interested in the intricacies of producing fine chocolates. All the social media accounts that Maggie had established directed viewers to the shop's proprietary website. After doing so, they could click a button and get a special discount for any orders placed online.

"Well, maybe using the social media accounts to drive business to our website for ordering is why it's been so busy lately," replied Maggie with a smile.

Assume you were Maggie. How could you use traffic tracking information to provide Otis with information about the impact of a social media presence on the increase in their online orders? How important would it be for you to do so?

Key Terms

Domain name	Proprietary website	Click-through rate (CTR)
IP address	NAP	Bounce rate
Extension (domain name)	Google search	Search engine results page (SERP)
Social media	Brand	Search engine optimization (SEO)
Registration (domain name)	F-shaped scanning pattern	Keywords
Web host	Navigation (website)	User-generated content (UGC) site
PDF	User experience	
	Traffic tracking (website)	

Operator's 10-Point Tactics for Success Checklist

Evaluate your need for, and the current status of, each of the following operational tactics. For those tactics you think are important, but not yet in place, develop an action plan for its implementation including who will be responsible for the tactic's completion and the target date by which it should be completed.

Tactic	Don't Agree (Not Done)	Agree (Done)	Agree (Not Done)	If Not Done Who Is Responsible?	Target Completion Date
1) Operator understands the importance of choosing a primary domain name that will drive traffic to the operator's proprietary website.	——	——	——		
2) Operator has carefully considered the important factors to be evaluated when choosing a website host.	——	——	——		
3) Operator understands the importance of having a professionally developed proprietary website.	——	——	——		
4) Operator understands the importance of professional website design when creating a proprietary website.	——	——	——		
5) Operator understands the importance of including key website content when creating a proprietary website.	——	——	——		
6) Operator understands the importance of evaluating tracking traffic on their operation's proprietary website.	——	——	——		

(Continued)

				If Not Done	
Tactic	**Don't Agree (Not Done)**	**Agree (Done)**	**Agree (Not Done)**	**Who Is Responsible?**	**Target Completion Date**
7) Operator has a basic understanding of how a search engine results page (SERP) is generated.	——	——	——		
8) Operator understands the importance of search engine optimization (SEO) efforts related to their proprietary website.	——	——	——		
9) Operator recognizes the increasing importance of social media in their overall marketing efforts.	——	——	——		
10) Operator understands the value of regularly posting and updating content on their chosen social media site accounts.	——	——	——		

5

Web-Based Marketing on Third-Party Operated Sites

What You Will Learn

1) The Importance of Marketing on Third-party Websites
2) The Importance of Marketing on User-Generated Content (UGC) Sites
3) How to Positively Impact Scores on User-Generated Content (UGC) Review Sites
4) How to Respond to Negative User-Generated Content (UGC) Site Reviews
5) How to Evaluate Marketing Efforts

Operator's Brief

In the previous chapter, you learned that a foodservice operator's proprietary website is most often its single most important marketing tool. In this chapter, you will learn how content from your proprietary website can be viewed on the websites of other entities.

There are several reasons why connecting (linking) your own website with third-party operated websites can assist your efforts to operate your business successfully. Perhaps most importantly, when your website is linked to those of other popular sites, visitors to those websites can click on a link and be directed to your website. In this chapter, you will learn how to establish linking relationships with desirable third-party website operators and some advantages of connecting your operation's website to those sites.

Increasingly, some foodservice operators are creating partnerships with companies offering third-party food delivery services. There are advantages and disadvantages to these partnerships, and both are addressed in this chapter.

(Continued)

In this chapter, you will also learn that user-generated content (UGC) sites are among the most popular sites on the Internet. Those who view these sites are often heavily influenced by the reviews of foodservice operations they read. As a result, most foodservice operators want to increase the total number and quality of reviews posted online and, in this chapter, you will learn how to do this.

Sometimes the ratings posted online about an operation are below average. Then it is important for operators to take decisive steps to remedy the situation(s) causing these poor reviews. You should promptly and professionally respond to any negative user reviews of your operation that have been posted online and, in this chapter, you will learn how this is best done.

Regardless of the methods they use to attract guests, all foodservice operators must carefully evaluate the effectiveness of those methods. In the final section of this chapter, you will learn about a four-step process used to evaluate the results of your total marketing efforts. Then, by doing so, you will be able to improve the effectiveness of your future efforts.

CHAPTER OUTLINE

Marketing on Third-Party-Operated Websites
Partnering with Third-Party Delivery Apps
    Advantages of Third-Party Delivery App Partnerships
    Disadvantages of Third-Party Delivery App Partnerships
Importance of User-Generated Content (UGC) Sites
Improving Scores on User-Generated Content (UGC) Review Sites
    Increasing the Number of User Reviews
    Increasing the Scores on User Reviews
Responding to Negative User-Generated Content (UGC) Site Reviews
Evaluating Marketing Efforts
    Assessment of Marketing Strategies
    Assessment of Marketing Tactics

Marketing on Third-Party-Operated Websites

One good way for foodservice operators to expand the visibility of their proprietary websites is to link their own sites with other website operators in their local communities. Linking is accomplished by providing a third-party operated website's viewers with a **link** to a foodservice operation's own website. A link, short for "hyperlink," is simply a connection from one webpage to another.

Key Term

Link: An easy way to navigate from one webpage to another webpage, or from one section of a webpage to another section of the same webpage.

Links are used when website owners want to connect their webpages to that of a partner organization, client, friend, blogger, or hobby group. For foodservice operators with a professionally developed and effective proprietary website, links established with those who operate other websites can be an excellent way to expand their online presence and reputation.

The type of link typically of most importance to foodservice operators is a **local link** that connects the web content of their operation with that of another locally operated webpage. The purpose is to show that another organization with significant relevance in their local area trusts or endorses the operator's business. When

Key Term

Local link: Website links created to show that other entities with relevance in the local area trust or endorse a foodservice operation.

the visitors to a local website click on the link, they will immediately see the content presented on an operator's proprietary website because they will be transported to that site.

To illustrate how a local link can benefit both linking partners, assume that a foodservice operation is located within one mile of the city zoo. It is possible that some visitors to the zoo might look for a foodservice operation close to the zoo for food and/or beverages either before or after their visit to the zoo.

Assume, when reviewing the zoo's website to learn about operating hours and driving directions, potential visitors to the zoo also saw a link to the foodservice operation's webpage. Some of these potential zoo visitors might click on the operation's link to gain foodservice-related information, and that operation might gain potential customers.

As a second example, assume that a foodservice operation is located near a limited-service hotel that does not provide on-site foodservice options for guests. It is likely that the hotel's operator would want to create a local link with the foodservice operation to inform their hotel guests about convenient foodservice options in the local area. In return, the foodservice operation's link to the hotel's website could also help in giving directions to it (e.g. "We are located right across the street from the Comfort Inn on West Harrison Highway.").

Creating local links to popular websites may also help a foodservice operation improve its ranking on search engine results pages (SERPs; see Chapter 4). The primary purpose of a link is to increase the visibility of the operation's proprietary website, and there are several ways foodservice operators can do this:

1) Linking with other businesses

Effective link building is about building relationships, and businesses operating their own websites are more likely to add a link to a foodservice operator's website if they have a genuine connection to the operator.

Foodservice operators should research local businesses that provide products and services that complement those offered by the operator. For example, a foodservice operator may partner with a local taxi service by adding a link to the taxi company's site, and the taxi company could provide a link back to the operator's website. Both businesses would benefit from the referrals resulting from this linkage.

2) Sponsoring local events

In many cases, local community events may depend on sponsorships to cover the event's costs. In exchange for the sponsorships received, the event's organizers may provide promotional benefits including website links to the sponsor's business.

Foodservice operators can monitor their local areas for conferences, concerts, festivals, and sporting and professional events held within their target market area. Which are likely to attract people who are potential customers? The answer: Any event with significant numbers of attendees who might be part of a foodservice operation's target market may be a good fit for sponsorship.

3) Sponsoring local organizations

In most communities, there are large numbers of nonprofit organizations and charities that seek support from community business leaders. Sometimes, linking with these organizations is a better tactic than sponsoring a local event. The reason: Links with nonprofit organizations are often more continuous instead of lasting only a few days or weeks.

For many foodservice operations, logical links with local nonprofits include organizations whose goals address healthy eating and living, green operating practices, and providing food and shelter to those in need. Local school systems may also seek sponsorships for items and events for which external funding is needed. Examples include sponsorships addressing necessary teaching supplies, sporting events, and sponsoring music and arts programs.

4) Joining appropriate professional organizations

Nearly all cities offer operators several professional organizations including membership in a local area's Chamber of Commerce, Restaurant Association, or **Convention and Visitors Bureau (CVB)**. These opportunities allow operators to interact with other local professionals and, in many cases, a business can link to a membership directory page found on the organization's proprietary website.

Key Term

Convention and Visitors Bureau (CVB): The local entity responsible for promoting travel and tourism to and within a specifically designated geographic area.

In some cases, professional organizations may be based on demographic factors. For example, an operator may elect to support a Hispanic Chamber of Commerce, a Small Business Chamber, an LGBTQ Chamber, a Professional Women's Chamber of Commerce, and others.

5) Connecting with local bloggers

Most mid-size to larger communities have local bloggers who write about things to do in the area and/or about issues relevant to those living in the community. These blogs may provide foodservice operators with an opportunity to create mutually beneficial links. For example, an operator might invite bloggers to visit their operation and sample menu items at no charge. In exchange, the blogger might write and post a positive review and place a link to the operator's proprietary website on their blog. Additionally, an operator might offer a special promotion for anyone who places an order and states that they heard about the special promotion after first visiting the blogger's site.

6) Participating in local awards programs

In many communities, local media including radio, television, newspapers, and bloggers regularly conduct "Best of the Area" awards programs. These programs typically have many categories, but they almost always include foodservice operations in multiple categories (e.g. "Best Italian," "Best Pizza," "Best Food Truck," and others.)

High rankings in awards programs can result in a foodservice operation's website being promoted or listed and then linked with the website announcing the awards.

Find Out More

Link building is an important part of search engine optimization (SEO) efforts. The larger the number of high-quality sites that mention a foodservice operation's name and provide a link to its website, the higher the operation will rank in search engine results pages (SERPs).

While it does require some effort, in nearly all cases foodservice operators find they have opportunities to link their own sites with many others in their local communities.

To learn about some ways foodservice operators can improve their web presence by linking their websites with local organizations and businesses, enter "link building strategies for restaurants" in your favorite browser and view the results.

Foodservice operators can also place information from (and in many cases create links to) its proprietary website on other third-party websites. Note: unlike an operator's own website, third-party-operated websites control the content they want to include on their sites. In some cases, the content on third-party websites is supplied directly by an operator and, in other situations, this is not the case.

One example of a third-party content-controlled website providing information about a foodservice operation is a "Google" search. Chapter 4 addressed the fact

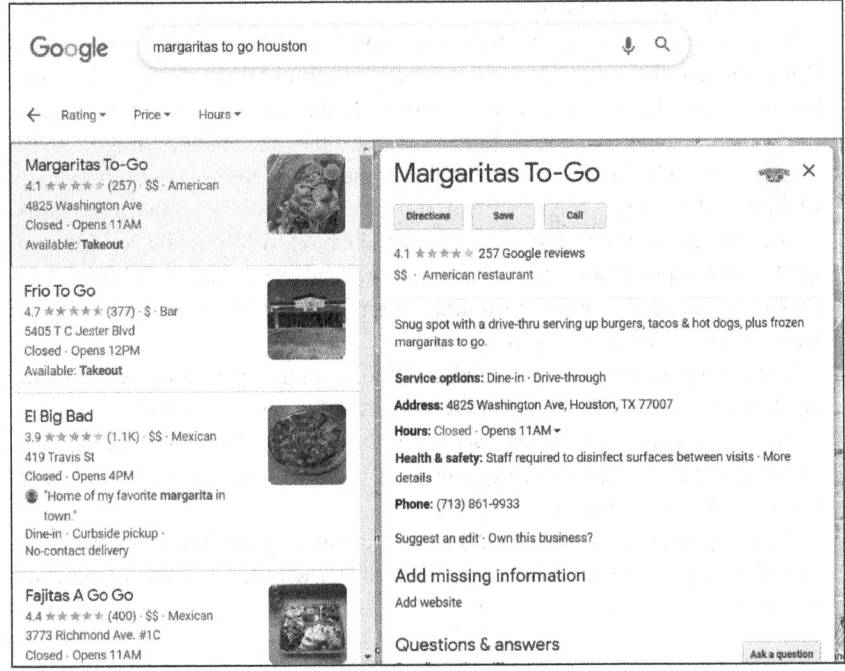

Figure 5.1 Google Search Results

that a SERP lists results returned by a search engine in response to a user's specific word or phrase query. Figure 5.1 shows a portion of the SERP produced when the foodservice operation "Margaritas To-Go Houston" is entered into a Google search.

The Google search results show Margaritas To-Go listed first on the SERP. In this example, Google has provided a significant amount of information about the operation including its NAP (see Chapter 4). Customer reviews have been posted along with the hours of operation, and space is provided for the operation to add its website as a link, if desired.

Note also that Google search has listed (left side of the results page), other food-service operations that might be of interest to the viewer/searcher.

The SERP Google generated for this search included, among many others, the additional third-party-operated websites listed in Figure 5.2.

The list above is not exhaustive, but it illustrates some of the different types of third-party operated websites listed on the SERP for this search. It is important to understand that all these sites (and others!) provide information about this foodservice operation. In some cases, the information provided on these third-party sites may be accurate and up-to-date. In other instances, they may not be because they are out of date, or an indi-cated operation may have closed. In all cases, the information will be displayed in a format determined by the third-party website operator.

https://www.facebook.com
https://365thingsinhouston.com
https://houston.eater.com
https://papercitymag.com
https://www.visithoustontexas.com

Figure 5.2 Selected Websites Displayed on "Margaritas To-Go Houston" SERP

Regardless of the information listed on the third-party site, a foodservice operator would not typically be allowed to go into these sites to correct erroneous information. Instead, the third-party website operator controls each site's content, and the operator may or may not provide an opportunity to easily correct or modify erroneous information. Therefore, it is important that foodservice operators routinely undertake Internet searches for their own businesses to learn what the SERP shows, and to then follow up with the third-party website operator to correct any out-of-date or otherwise incorrect information.

In many cases, foodservice operators can establish their own accounts on third-party operated social media sites. Facebook is a currently popular social media site, and it provides a good example of how a foodservice operation can create a Facebook account and use this social media site to its advantage.

Some content suggestions that could be placed on Facebook's third-party operated site (and other sites that permit it) include:

1) Posting menu item information and photos

For most foodservice operations, posting information on third-party operated websites is a good way to increase the visibility of the menu items they sell. Operators can easily take high-resolution photos of selected dishes they sell and publish them once or twice a week on their Facebook page. This search engine optimization (SEO) strategy works because SERPs are directly influenced by the frequency with which new information is placed on a website.

2) Showing the operation

Foodservice operations that offer pictures of on-site dining areas allow viewers on third-party operated websites to visit the restaurant virtually. Photos of the operation's exterior entrance doors, dining and lounge areas, and table settings can all increase viewers' comfort levels because they can see online what they will experience if they visit the operation.

3) Showing the operation's staff

With the permission of staff members, operators can post photos of their managers, servers, and kitchen staff on their Facebook pages. Pictures of smiling staff members wearing their uniforms can help put a personal "Face" on the operation.

4) Showing the operation's staff in action

Posting pictures of staff members serving happy guests or performing interesting kitchen tasks is typically a good idea for a Facebook posting. If operators utilize this marketing tactic, they must ensure that the staff members are demonstrating safe food practices and that any background areas shown in the photos are clean and professionally maintained.

5) Provide interesting food-related information

Guests viewing content posted on third-party operated sites are often interested in learning more about food and its preparation. Facebook postings can describe unique menu ingredients and where they are from, and/or unique preparation methods and why they are used. Both tactics help educate customers and enhance the personal relationship between the site's viewers and the operation.

6) Sharing positive guest reviews

Facebook pages allow guests to post reviews of the foodservice operations they have visited. When the reviews are positive, they can be highlighted and shared on the operation's initial Facebook page. These postings might include excerpts of positive reviews and a management reply thanking reviewers for sharing their opinion. Facebook page reviewers will also better understand that they will also receive a positive experience when visiting the operation.

Foodservice operators desiring to establish partnerships with third-party operated websites will find they have many opportunities to do so. Each third-party website operator may have their own rules and procedures for posting content. Foodservice professionals should select the third-party operated sites they feel will be most useful to their operations. They can then make maintaining a relationship with those sites an important part of their annual marketing efforts. These efforts will typically include keeping their accounts active and, in some cases, deciding to advertise on the sites.

Technology at Work

The specific ways that social media sites permit foodservice operators to advertise often varies based on the site. The specific form of an ad to be posted on a social media site is also most often best determined based on the site itself. For example, sites featuring video ads such as YouTube and TikTok typically are the most effective because visitors to these sites expect to see videos. In contrast, other sites print advertisements that may (or may not) include photographs.

Navigating all these social media sites and their varied advertising forms, requirements, and benefits can be overwhelming for some foodservice operators. This is also true about the effectiveness of various metrics (measures of quantitative assessment commonly used for comparing, and tracking performance) that site operators provide to their advertisers. Fortunately, there are numerous companies to assist advertisers as they create and post effective social media ads.

To review some of these companies and the services they provide, enter "Social media advertising assistance for restaurants" in your favorite search engine and view the results.

Partnering with Third-Party Delivery Apps

Foodservice operators make important business decisions as they consider whether to partner with one or more businesses operating **third-party delivery** apps. Companies that have created these apps can make it easy for foodservice guests to place online orders for meal delivery to their homes, offices, or other desired locations.

The COVID pandemic of 2019 significantly increased the use of third-party delivery apps because many foodservice operations were closed for indoor dining and, also, many guests decided it was safer to eat in their homes than to dine out. Today DoorDash, Uber Eats, Postmates, and Grubhub are among the largest and most popular third-party delivery companies that have developed apps. Each of these companies exhibits strengths in some geographic markets but have a lesser impact in other areas.

The basic business model of third-party delivery companies is relatively straightforward. Foodservice operators agree to have their operations listed on the app, and the operator then provides a menu that includes item prices. Guests utilizing the apps place their orders with the third-party delivery company (they do not interact directly with the operation).

Orders placed on the app are electronically transmitted to the foodservice operation's point-of-sale (POS) system. The third-party delivery company uses a cadre of independent drivers to pick up the orders and deliver them to guests. In addition to paying for the food they have ordered; customers also pay a delivery fee to the third-party delivery company. Drivers are paid on a per order basis by the company, and they typically retain any tips generated by the orders they have picked-up and delivered.

The third-party delivery company charges a commission to the foodservice operation for its services. Typical commission fees range from 20% to 30% of the dollar amount of the order. The remaining amount of the food order's value is then transferred to the foodservice operation's bank account.

Key Term

Third-party delivery: A smart phone or computer application that creates a marketplace that customers can search to browse restaurant menus, place orders, and have them delivered to a location of the customer's choosing. In nearly all cases the guest orders are delivered by independent contractors who have been retained by the company operating the third-party delivery app.

Advantages of Third-Party Delivery App Partnerships

Unless required as part of their franchise agreement as negotiated by the franchisor, foodservice operators have the discretion to create partnerships with third-party delivery organizations. The advantages of such partnerships for a third-party delivery company are easy to understand. The more foodservice operators listed on the third-party site, the better the chances that customers will find an operation from which they wish to order and the higher will be the commissions earned by that company.

As they market apps to foodservice operators, third-party delivery companies generally stress three key factors to encourage foodservice operation sign-ups:

✓ App popularity
✓ Competitive pressures
✓ Impact on profits

App Popularity

Online ordering is increasingly popular among guests. Online food ordering has grown 300% faster than dine-in since 2014 and now accounts for roughly 40% of the total restaurant sales. In fact, 60% of Americans order delivery at least once per week.[1]

Third-party delivery app operators point to the increasing popularity of home meal delivery as one reason foodservice operators should join them. Note: this reasoning is often generated because of the above-cited data related to changes in the market behavior of today's consumers.

Competitive Pressures

A second major reason frequently cited by third-party app operators seeking partnerships is that a foodservice operation's direct competitors will be listed on their sites. As a result, some foodservice operators reason that, since their competitors will be listed on the site, they must also be listed to maintain competitiveness.

Using this argument, third-party app operators point to an increasing number of customers who desire online ordering and delivery option(s). If this service is not offered, some potential customers will select a different restaurant that does offer delivery, and the operator will lose the chance to benefit from these customers' sales.

Impact on Profits

Foodservice operators who partner with third-party delivery services typically do so with the belief their profitability will increase. In fact, third-party delivery companies often indicate that cost savings do accrue to foodservice operations that do not need to create a unique online order and delivery system on their own websites because these are provided by the third-party service.

---

1 https://www.zippia.com/advice/food-delivery-industry-statistics/ retrieved October 21, 2023.

Third-party delivery app operators also suggest that the typical online delivery order is larger than that typically placed by on-site diners (the guest check average is higher with online delivery services). Also, since third-party delivery operators charge commissions on each completed sale, the foodservice operation only pays a fee when a sale is generated.

Disadvantages of Third-Party Delivery App Partnerships

Despite their increasing popularity, some foodservice operators point to perceived disadvantages of developing partnerships with third-party delivery companies. While the precise reasons vary, operators who do not strongly support third-party delivery app partnerships typically point to three areas of concern:

✓ Quality control
✓ Guest ownership
✓ Financial concerns

Quality Control

When partnering with third-party app delivery companies, foodservice operators have 100% control over ordered menu items as they are produced and packaged for delivery. Once the food is picked-up for delivery, however, operators have virtually no control if challenges arise.

For example, if it takes a long time for a driver to deliver an order, there is little the foodservice operator can do. Similarly, if a driver's appearance is unprofessional, they "sample" the guest's order, smokes, or makes deliveries with pets in their vehicles, the third-party service (not the operator) must address these or related issues. Given the critical importance of guest ratings posted online, poor reviews resulting from improper delivery service, rather than poor menu item preparation quality can be very damaging to a foodservice operator's reputation.

Guest Ownership

Guest ownership is a significant issue for operators who are wary of third-party delivery apps. If a guest uses an in-house operated online ordering system or personally calls in an order, the customer can be enrolled in and earns points from the operation's **customer loyalty program**. This, in turn,

Key Term

Customer loyalty program: A marketing tool that recognizes and rewards customers who make purchases on a recurring basis. The programs typically award points, discounts, or other benefits that increase as the total amount of a repeat customer's purchases increase.

Also commonly known as a "Frequent guest program," "Frequent customer program," or "Frequent dining program."

encourages repeat business, and information about the customer such as credit card data, e-mail address, and a telephone number could be retained by the operation and used for other marketing purposes.

Guests utilizing a third-party delivery app, however, are technically customers of the third-party delivery app company. As a result, vital guest information including size of order, frequency of ordering, and preferred payment method will be retained by the third-party app, and this important data will not likely be shared with the foodservice operator.

Financial Concerns

One complaint of foodservice operators (and some guests!) about third-party delivery app operators involves the delivery and service fees that must be paid. For foodservice operators, the delivery fees paid to third-party delivery app personnel reduce the amount of money remaining to pay for food, labor, and other operating expenses required to produce the menu items delivered. In many cases, after delivery fees are considered, foodservice operators often make very small profit margins, may break-even, or may even lose money on a delivery sale.

Given the potential advantages and disadvantages of partnering with third-party delivery apps, foodservice operators generating a significant amount of delivery business should also consider the advantages and disadvantages of implementing a self-delivery program. While there are costs related to doing so, there might be significant advantages to implementing a self-delivery, and they include:

✓ Quality control of delivered food items
✓ Ability to establish delivery fees
✓ Use of branded uniforms and transportation vehicles
✓ Increased control of guest data
✓ Ability to interface with the property's existing guest loyalty program
✓ Elimination of commission and fees
✓ The possibility of using current staff for deliveries during slow periods
✓ The creation of jobs
✓ Increased profits

The decision to partner with third-party delivery app companies is an important one, and the concept must be carefully considered by foodservice operators before they elect to engage in such a partnership.

Find Out More
Domino's Pizza, headquartered in Ann Arbor, Michigan, is the world's largest pizza chain. As of 2023, the company franchised or operated more than 14,400 stores in 85 countries.
Pizza is the company's primary focus with traditional, specialty, and custom pizzas available in a variety of crust styles and toppings. While menus vary

somewhat by location, in most stores additional entrees include chicken wings, pasta, bread bowls, and oven-baked sandwiches.

Domino's promises fast delivery and at one point in its history (1973), the organization guaranteed that guest orders would be delivered in 30 minutes or less (a practice it has since discontinued due to safety concerns). The company is innovative in numerous ways and, in 2021, it began testing self-driving delivery vehicles. In fact, Domino's does *not* now, nor has it ever, partnered with third-party delivery companies. Instead, the company hires delivery employees and supplies necessary delivery vehicles.

For operators considering launching or continuing their own delivery service, the Domino's perspectives on third-party delivery partnerships may be enlightening. To learn more about this approach to home meal delivery, enter "Domino's home delivery innovations" in your favorite browser and view the results.

*https://www.dominos.com/en/about-pizza/pizzeria/ retrieved October 22, 2023.

What Would You Do? 5.1

"What did the salesperson say?" asked Malik, the assistant manager of the On-Campus Deli.

Malik was talking to Veronica Richards, the owner of the On-Campus Deli, a popular New York-style delicatessen located adjacent to a major college campus. Veronica's restaurant featured traditional deli fare including corned beef and pastrami sandwiches, homestyle soups, and traditional cheesecakes. Veronica employed a small number of delivery employees who delivered call-in and online orders to students and faculty living in or near the college. Business was good, and orders placed for delivery were increasing significantly. For that reason, Veronica had met with a sales representative for Diners' Dash, a third-party delivery company offering its services to restaurants in the local area.

"She said that, with our increasing demand for off-site delivery, it is a good idea to partner with them," replied Veronica.

"That makes sense to me," said Malik, "what do you think?"

"I don't know. I think we should keep our own drivers and work to decrease demand for off-site delivery. I think we can do that by offering more margin-friendly curbside and takeout options to our guests. If we reduce the prices for guests coming to get their own food, we'll be moving the amount we would have paid to the delivery company back to our own guests. I think I'd like to try that approach first," replied Veronica.

Assume you were Veronica. What would be some risks associated with seeking to de-emphasize the off-site delivery of guest orders? What could be potential advantages of implementing such a strategy?

Importance of User-Generated Content (UGC) Sites

In Chapter 4, a user-generated content (UGC) site was described as any social media website in which most content including images, videos, text, and audio was posted online by the site's visitors. Today there are numerous popular UGC sites including Instagram, Facebook, Pinterest, TikTok, Snapchat, X (formerly Twitter), and YouTube. Today, the most read and viewed content on the Internet is that contributed by visitors to UGC sites. The latest statistics show that Facebook had 3.03 billion active users in 2023.[2] That means that nearly two out of every three of the 4.89 billion social media users across the world are active users of Facebook. The amount of information posted online by users is extraordinary.

Experienced foodservice operators recognize the power of UGC sites because they understand that people more readily trust the recommendations of a "real" person than what a business says about itself. For example, a recent study indicated that 56% of consumers said they more readily trust user-generated content and that it was what they most wanted to see from businesses. Note: only 15% of the study's consumers said content created by businesses was what they wanted to see.[3]

Since the popularity of specific UGC social media sites will likely continue to rise and fall and new sites will be introduced, their importance to foodservice operators as they market their businesses cannot be overstated. This statement is true in terms of their use as an advertising channel, and also the importance of guest reviews and ratings posted on these sites.

It is not usually necessary for a foodservice operation to have an active social media account on a UGC site and then to advertise on it. For example, a foodservice operator may advertise on Facebook or LinkedIn without creating and maintaining an active Facebook page or LinkedIn account. Doing so is often called **paid social advertising**.

Key Term

Paid social advertising: The use of ads on social media sites for which an operator must pay a fee.

Each social media site has its own method of charging for the advertisements it displays. In some cases, the fee is a flat dollar amount based on an ad's size and the amount of time the ad is displayed. In others, the fee is based on the number of times the advertisement is accessed (clicked on) by social media users.

---

2 https://www.shopify.com/blog/most-popular-social-media-platforms#:~:text=The%20latest%20statistics%20show%20that,are%20active%20users%20of%20Facebook retrieved October 22, 2023.
3 https://www.nosto.com/blog/what-is-user-generated-content/#:~:text=UGC%20is%20authentic&text=People%20more%20readily%20trust%20the,what%20they%20wanted%20to%20see retrieved October 22, 2023.

Regardless of the social media sites on which they prefer to advertise, foodservice operators can employ several key actions as they plan and implement their paid social efforts:

1) Define the target audience
2) Create memorable content
3) Change ads frequently
4) Analyze the data produced

Operators of social media sites can provide significant amounts of an ad's effectiveness-related data. This should be regularly and carefully reviewed and analyzed to determine if the impact of the advertisement is positive.

Improving Scores on User-Generated Content (UGC) Review Sites

Many popular UGC sites enable the site's users to post reviews of the restaurants they have visited (some UGC sites significantly feature restaurant reviews or are even devoted almost exclusively to these reviews!). While the popularity of any given UGC review site will likely increase or decrease over time, some of the most popular review sites include:

✓ Yelp
✓ OpenTable
✓ TripAdvisor
✓ Zagat
✓ Zamato (formerly Urbanspoon)
✓ Google My Business
✓ Facebook
✓ Reddit
✓ Third-party delivery app sites

Note that this list does not include popular third-party delivery app sites (e.g. Grubhub, DoorDash, and Uber Eats) but sites such as these also offer a user-review option.

Marketing research has shown that most reviewers want to help the business by leaving a favorable review. If foodservice guests have a positive experience with an operation, they want to let others know about it so review readers can have the same enjoyable experience. Note: Most reviewers also believe they are rewarding the company for the excellent food and/or service they received.

When negative reviews are written it is likely that a guest had one or more problems that were not corrected on-site or upon (or after) a product delivery. This demonstrates that problems solved immediately lead to higher online scores because guests with corrected negative experiences are less likely to write about the problem if it was resolved to their satisfaction.

Potential guests pay a great deal of attention to reviews posted online. Foodservice operators want to have as many positive reviews posted about their operation as possible. High scores on customer reviews are the result of two specific management activities:

1) Increasing the number of user reviews
2) Increasing the quality of user reviews

Increasing the Number of User Reviews

Online guest reviews are today's equivalent of "yesterday's" word-of-mouth advertising. Traditionally, those desiring information about a foodservice operation asked friends or family members for recommendations. Today, Internet users turn to online customer reviews to help determine the operations they will visit for the first time.

For the majority of UGC sites featuring customer reviews, three factors determine how high an operation will be listed on its search engine results page (SERP; see Chapter 4). These are:

✓ Number of reviews (more reviews are better than fewer reviews)
✓ Quality (scores) of reviews (good reviews are better than poor reviews)
✓ Recency of reviews (recent reviews are better than older reviews)

Foodservice operators can employ several strategies to increase the number of reviews their operation receives:

1) Featuring positive reviews

A foodservice operation's proprietary website was described earlier as an operation's single most important online communication tool. Therefore, featuring positive reviews on the home page of an operation's proprietary website is always a good idea. These reviews can be posted in their entirety, or they can be highlighted with the use of a specific "Reviews" tab that, when clicked on, displays positive reviews of the operation. This approach can help increase the number of reviews because it encourages others to post their own additional reviews because they know those reviews will be valued and may be showcased by the operation.

2) Linking to popular third-party social media sites

When foodservice operators have created their own accounts on popular social media sites, the sites should be linked to the operators' proprietary websites. For example, it should be easy for a website viewer to click on a Facebook link placed on the proprietary site and then be taken to the operation's Facebook page where reviews can be posted. Generally, foodservice operators should create accounts on the social media sites their guests are most likely to utilize.

3) Asking current customers to give a review

In many cases, foodservice personnel can simply ask guests at the end of their meal if they would be willing to review the property. This can be done in a nonintimidating way by servers who simply say, "If you enjoyed your meal with us today, we'd love for you to leave a review for us!" In many cases, this will result in guests leaving a positive review on the social media UGC sites they themselves use.

4) Rewarding those who leave a review

One good way to reward those contributing a review is by offering something such as a 10% off coupon for future purchases. This strategy can be employed even if the discount is not disclosed to the guest before the review is written. In that scenario, it is highly likely the guest will be delighted when they receive the unexpected discount and may even tell others about it (which may encourage even more positive reviews!).

5) Making it easy to leave a review

Perhaps the most important thing foodservice operators can do to increase the number of their reviews is to make it easy to post a review. To do this, it is important for foodservice operators to understand the review posting process. To illustrate, consider the process used to post a Google review. Note: Google is an extremely popular UGC site that features restaurant reviews that follow several steps:

1) Open Google maps
2) Search for the foodservice operation's name
3) Click on the operation's name to pull up its Google business profile
4) Scroll down to the review section
5) Click on "Write a review"

Now, consider the situation in which a guest who just finished a meal wants to depart and says to an operation's manager, "My meal was wonderful! How can I leave a review?" The manager's response in this scenario could be:

"Thanks very much for asking about the review process, and yes you can leave a Google review. It's easy. Just go to Google Maps and do a search for us by entering our name in the search bar. When you pull up our listing, scroll down the page until you come to the review section. Then just click that button, and you can write a review"

Or the response could be:

> *"Thanks. Just go to our website and click on the Google Review link button located on our homepage!"*

Clearly, the operator using the second response is more likely to get a review! Foodservice operators should provide direct links to review sites that are important to them as well as to those that are currently of most importance to their target markets.

Increasing the Scores on User Reviews

The score reported on a UGC site featuring guest reviews allows the reviewer to rank the quality of the experience that occurred when the reviewer interacted with the operation. When foodservice operations have met or exceeded these guests' expectations, they are highly ranked. In contrast, lower rankings are given when the reviewer's expectations were not met.

Various UGC sites featuring user reviews have different scoring systems. For example, on Yelp, scores can range between 1 and 4 "Stars." On Google and Facebook, the scores range between 1 and 5 "Stars." On TripAdvisor, reviewers assign a score of 1–5 that results in the display of filled or partially filled "Circles."

Grubhub also uses a 5 "star" rating system. It also publishes an additional summary, "Here's what people are saying," about a business. For example, the review for a single operation might say, "Here's what people are saying: 93% Food was good, 88% Delivery was on time, 82% Order was correct."

On Google and TripAdvisor, among others, the sites' operators also use algorithms to rank operations. For example, a search for "Best pizza near me" will produce a SERP that ranks pizza restaurants "in rank order." However, the ranking includes more factors than just the average reviewer's score: frequency and recency of reviews are also factored into the ranking.

While the specific scoring and ranking systems used by UGC review site operators vary, the most essential fact foodservice operators must recognize is that the scores matter a lot. Online market research indicates that 70% or more of site visitors reviewing online ratings only continue to read information about an operation if its ratings are within the top two scores (3 and 4 for a four-point rating system, and 4 and 5 for a five-point rating system).[4] Therefore, scores achieved by an operation directly affect the amount of business they will do. In fact, a comprehensive study by Professor Michael Luca at the Harvard Business School found that a 0.5-star increase in an average Yelp rating generates a 9% revenue increase in revenue for an independent restaurant.[5]

---

4 https://www.qualtrics.com/blog/online-review-stats/ retrieved November 20, 2023.
5 https://www.hbs.edu/ris/Publication%20Files/12-016_a7e4a5a2-03f9-490d-b093-8f951238dba2.pdf retrieved October 23, 2023.

Since UGC review scores do matter, proactive operators can employ several strategies to increase the average review score their operations receive:

1) Monitoring the reviews

Foodservice operators monitoring their online reviews can learn what their customers think about them. Certainly, online reviews communicate to potential guests who read them, but they also provide important information to foodservice operators who want to improve their scores.

Foodservice operators can often gain insight into their own businesses by reading their reviews. For example, assume that a consistently high number of reviewers indicate that parking when arriving to pick up their orders is a problem for guests. In this example, the operation is clearly being told that it needs to take steps to improve its product pick-up system. This may entail, for example, designating special parking spaces for pick-up orders or providing servers who can deliver orders to guests' arriving vehicles. Similarly, if review scores posted on third-party delivery app sites are low, it may indicate a problem with third-party delivery order preparation, packaging, or delivery to drivers.

Another key reason for monitoring reviews is to measure an operation's success in improving its scores. Assume a foodservice operation was achieving a score of 4.2 stars on a popular user review site. Assume further that the operator wanted to increase that scoring average to 4.5 in the coming marketing period. In this example, monitoring score movement would be important for the operator to determine whether the marketing strategy and tactics employed in this area are meeting with success.

2) Encouraging reviews from satisfied customers

SERP results are weighted-based on the number and recency of reviews, but they are most heavily influenced by the actual scores given by reviewers who submit a rating.

UGC sites featuring user reviews use an **arithmetic average** to calculate the average rating of a business. The more positive reviews received, the higher the average rating will be. It is always helpful to encourage reviews from guests who are highly satisfied with their experience.

3) Fixing problems when they occur

An arithmetic average is reduced when review scores are low, so low review scores should be avoided, when possible. The best way to do this is by ensuring that any problems that

Key Term

Arithmetic average: The simplest and most widely used measure of a mean (average); it is calculated by dividing the sum of a group of numbers by the count of the numbers used in the series. For example, if the reviewers' scores on a review site are 3, 4, 4, 5, and 5, and the sum is 21, the arithmetic mean is divided by five and equals 4.2 (21/5 = 4.2).

occur during a guest's visit are immediately resolved *before* the guest leaves the operation.

Problems with food quality or service delivery can occur in any foodservice operation. When operators ask about guests' experiences, those with a negative experience often have their problems immediately corrected. Doing so greatly reduces the chances that the guest will post a negative online review and may even result in a positive review describing how the operation corrects its mistake.

4) Responding to reviews

Recent reviews *and* recent responses weigh heavily on SERP results. A foodservice operator receiving dozens (or even hundreds!) of reviews need not respond to every review, and the importance of professionally responding to all negative reviews is addressed later in this chapter.

Operators should regularly (several times weekly, if possible) respond to online reviews. Creating a dialogue between the operator and customers is perceived positively by those reading online reviews. Responses to reviewers also indicate to UGC site visitors that the business considers and values the quality of its customers' experiences as the operation is managed.

The responses to a positive guest review need not be long and detailed. A simple "thank you" and an expression of pleasure that the guests received a positive experience typically suffice.

5) Providing excellence in food quality and service

Guests selecting a foodservice operation anticipate they will have a good experience, and they are typically predisposed to think their decision to visit an operation was a good one. Providing excellence in food quality and service is the best way to verify that the decision was, in fact, a good one!

An operation that works hard to provide high-quality guest experiences is likely to avoid many poor reviews. Paying close attention to food and service details encourages high review scores because doing so results in satisfied customers. These review scores will also "invite" potential guests reading high-ranking online reviews to visit in the future.

Responding to Negative User-Generated Content (UGC) Site Reviews

Even the best-managed foodservice operations will encounter some negative reviews posted on UGC review sites, and operators should respond to these reviews promptly and professionally for several reasons:

1) A boost in ranking or listing: Many data integration programs are developed to respond to and reward business-to-consumer interactions. When these

programs detect that a business is regularly communicating with its customers, the business will be placed higher on a SERP page.

2) Showing the business cares: When customers complain, they typically want to know that someone hears them. A response (of almost any kind!) demonstrates that management does care about its guests' challenges and is interested in addressing them.

3) Minimizing the damage: Once a legitimate negative review of an operation has been posted online, it will not, in most cases, be taken down at the request of the operator. However, when readers or viewers of a negative review immediately encounter a sincere and timely professional management response, the reputational damage done by that negative review is lessened.

4) Demonstrating professionalism: A rapid response to negative reviews shows that a business is concerned about its customers, is responsive to them, and will respond quickly and professionally when issues arise. These are important traits all consumers look for when they buy products or services.

Foodservice operators do not have the luxury of being able to ignore negative reviews in the hopes they will go away because, in nearly all cases, they will not do so. Ignoring negative reviews increases their impact because a reply is not just a response to that person. Instead, it is a message to all others who will read or see the review including potential guests.

A lack of response by a business tends to confirm the general position of the reviewer (the business does not care about its customers!). Figure 5.3 lists guidelines for how foodservice operators can positively respond to negative online postings about their business.

---

1) Thank the reviewers for taking time to give feedback. (It is always best to start a response in a positive way.)
2) Apologize if the customer is correct about what happened.
3) Avoid arguing if the customer is incorrect about what happened. Instead, apologize and indicate that there was a misunderstanding.
4) Provide a very brief but clear and direct explanation of what caused the problem.
5) Assure the site's readers that specific actions have been taken to avoid a repeat of the problem.
6) Offer a direct line of communication between the business and the negative reviewer via e-mail, text, or phone to receive a personalized apology.
7) Conclude the response by directly quoting any possible part of the reviewer's comments that were positive (e.g. if the reviewer says the server was "friendly" but the soup was "cold," mention that it was good to learn the reviewer found the server to be "friendly.")
8) End the posting by *again* thanking the reviewer for feedback and helping to make the operation even better because of it.
9) Invite the customer to give the operation another chance to provide excellent service and products.

Figure 5.3 Guidelines for Responding to Negative Online Reviews

Note: when responding to negative and positive reviews, operators must always remember that their responses should be well-written and grammatically correct.

Technology at Work

Foodservice professionals in all segments of the hospitality industry understand that the world is more connected than ever due to the wide reach of the Internet. Online reviews are tools consumers use to communicate with each other about the quality of the foodservice operations they have visited. While most reviews posted online will be good ones, negative comments about a business or brand can create significant damage. Online review of monitoring companies can offer significant value to foodservice operators of all sizes, and they have programs to continually scan the Internet to discover positive and negative reviews about their clients.

Sometimes consumers post unfair or untrue reviews. If this occurs, an online review monitoring company can often help mitigate them. This will protect an operation's reputation because the company understands the required methods to help operators remove fake reviews from a site. The online review monitoring company can also assist in identifying positive reviews and disseminating them as widely as possible.

To examine some of the online review monitoring companies and to learn about the services they can provide, enter "Assistance in monitoring online restaurant reviews" in your favorite search engine and view the results.

Evaluating Marketing Efforts

The efforts used to market a foodservice operation are many and varied. As addressed in Chapter 3, these efforts are formalized in an operation's marketing plan that indicates the marketing efforts to be undertaken, when they will be undertaken, and who will complete them.

When the time covered by the marketing plan has passed (and regularly thereafter) effective foodservice operators evaluate the results of their marketing efforts. This involves assessing the extent to which strategies (goals) identified in the marketing plan were addressed and an evaluation of the tactics used to attain them.

Several factors can impact the results obtained by an operation's marketing efforts. For example, an operation may have a very effective marketing program. However, if it fails to deliver on its promises when guests visit the operation, long-term desired financial results may not be achieved. Alternatively, an operation may do only a limited amount of marketing, and its guest counts might continually increase because the product quality and service levels of the operation are consistently high.

The reasons for undertaking a detailed examination of an operation's marketing results are many and include:

✓ Gaining a better understanding of a target market's behavior

✓ Allowing for a quantitative assessment of key **revenue metrics** that are impacted by marketing strategies and tactics

✓ Providing a systematic assessment of online reputation and brand awareness

✓ Making better decisions for future marketing efforts

Key Term

Revenue metric: A standard for measuring or evaluating data based on its dollar amount or change in dollar amount. Also known as a "financial metric."

Internal and external factors can impact an operation's ability to achieve marketing plan objectives. A regularly planned and objective assessment of marketing efforts is required so operators can identify marketing activities that worked well and others that did not work as well. This information is essential if managers are to improve their marketing efforts from one time period to the next.

As shown in Figure 5.4, there are four important cyclical steps needed to assess an operation's marketing efforts. The operator must (first step) assess the current marketing plan results and, (second step), modify marketing strategies and tactics as needed. It is then possible to prepare (Step 3) and implement (Step 4) the next period's marketing plan. At some point that marketing plan will again be revised, and the cycle in Figure 5.4 will be repeated.

Assessment of Marketing Strategies

Chapter 3 defined a marketing strategy (goal) as an objective a foodservice operation wishes to achieve, and the best of these objectives are both realistic and measurable. Likewise, when they assess the achievement of their market strategies, operators compare planned results with actual results.

For example, assume an operator identified an objective to increase pick-up business by 10% during the time span of a marketing plan. That objective can be quantifiably measured and, when the time addressed in the marketing plan is concluded, the degree to which the objective was achieved can also be assessed. If the objective was achieved, the operator can look to the specific tactics that were successfully used to support the objective. If the objective was not achieved, the foodservice operator can then assess the individual tactics put in place to help

Figure 5.4 Marketing Efforts Assessment and Improvement

achieve the objective. The operator can also judge which tactics were successful and identify others that were not effective and need replacement or revision.

There are numerous external factors that can contribute to a foodservice operation not achieving marketing goals. A depressed economy, layoffs by a major employer in an operation's marketing area, and nearby road construction are examples. Others, including the opening of a competitive operation and severe weather such as hurricanes or floods, can also impact an operation's revenue-generating abilities and opportunities to meet marketing goals. As they assess marketing success, operators must keep external factors in mind and make a reasonable determination about reasons for unattained objectives.

Assessment of Marketing Tactics

Marketing tactics are the specific steps and actions operators undertake to achieve their marketing objectives. An assessment of marketing tactics is useful regardless of whether the planned objectives were or were not achieved.

When assessing specific marketing tactics, operators should consider the answers to three important questions:

1) Was the tactic implemented?

Foodservice operators must identify tactics to be implemented that support their marketing goals. In some cases, a tactic is not fully implemented during the time covered by the marketing plan. For example, assume a foodservice operator identified implementing an **up-selling** training program as an important tactic that would attain a 5% increase in guest check average during the time addressed by the marketing plan.

Key Term

Up-selling: The process of increasing the guest check average through the effective use of on-site selling techniques; also referred to as "suggestive selling."

The marketing plan called for the up-selling training to occur during the first quarter of the year. In fact, the training was not implemented until the third quarter of the year. In this example, while the tactic was implemented, the timing of the training did not optimize its effectiveness in helping the operation to achieve a 5% increase in guest check average.

2) Was the tactic cost-effective?

As marketing funds are expended, foodservice operators know the costs of implementing specific tactics included in their marketing plans. In some cases, the actual cost of implementing the tactic is consistent with the original budget forecasts. In other cases, however, the cost of implementing the tactic may be more (or less than) originally forecast.

In many cases, foodservice operators undertake specific marketing tactics to increase guest counts, increase total revenue, and/or increase guest check average. Assessing the actual cost of implementing a tactic is important to determine the tactic's **marketing return on investment (ROI)**.

Key Term

Marketing return on investment (ROI): The amount of benefit gained from a marketing activity compared to the cost of undertaking and implementing the activity.

A marketing activity's ROI is the difference between how much was spent on a specific marketing activity and how much revenue it generates. The formula foodservice operators use to calculate their marketing ROI is:

$$\frac{\text{Marketing revenue} - \text{marketing cost}}{\text{Marketing cost}} \times 100 = \text{Marketing ROI}$$

To illustrate, assume an operation implements a marketing tactic that cost the operation $28,000. The impact of the tactic was an increase in revenue of $98,000 during the time covered by the marketing plan. In this example, the operator's ROI for this marketing tactic would be calculated as:

$$\$98,000 \text{ revenue} - \$28,000 \text{ marketing cost} = \$70,000$$
$$\$70,000/\$28,000 = 2.5$$
$$2.5 \times 100 = 250\%$$

In this example, the operation achieved a 250% ROI on its marketing activity. Stated another way, for every dollar the operation spent implementing this marketing tactic, $2.50 in additional revenue was generated.

The measurement of ROI allows operators to compare the revenue impact of various marketing tactics. For example, assume that an operator finds that, when an investment is made in online marketing efforts, the return on investment is 300%. When the organization invests in local newspaper ads, the resulting ROI is 125%. In this example, it may make good sense for the operator to increase the amount of online advertising and reduce the amount of paid newspaper advertising because of the resulting ROIs.

3) Was the tactic results-effective?

In many respects, this is the most important question of all, and it can also be one of the most challenging questions for an operator to answer. In some cases, operators can identify quantitative measures of advertising effectiveness. For example, they can monitor total revenue generation, count the number of new customers signing up for guest loyalty programs, measure total website traffic, and/or assess website click-through rates (see CTRs in Chapter 4).

While each operation is different, in many cases advertising effectiveness can be assessed by:

Monitoring Guest Counts

One relatively easy way to determine if advertising is working is to record changes in guest traffic by counting the number of guests who are served by the foodservice operation or who visit a website. When operators monitor traffic before they start an ad campaign, they will then have a basis for comparison. It is also helpful to ask new customers how they heard about the business.

Coding Coupons

When possible, producing ads and promotions with a coupon that customers can redeem for a purchase discount is a good way to measure impact. When the coupons are coded with a unique identifier, an operator can determine which ad, website, or print publication generated the best results.

Offering Incentives

Foodservice operators can offer incentives for guests in their advertising efforts. For example, an operation may offer incentives for customers to communicate they are responding to an ad by stating "Mention this ad and get a 10% discount on your meal purchase." When incentives are advertised on a website, in local publications, or on radio the operators will know the impact of the advertising medium based on the number of customer responses to the incentive.

Monitoring Sales

Monitoring sales means making careful assessments before, during, and after an advertising tactic is undertaken. What foodservice operators must recognize, however, is that advertising may have a cumulative or delayed effect, and advertisement-driven sales may not be identified immediately. This is especially so when an advertising tactic is designed to support a goal of increasing brand awareness or enhancing the reputation of an operation in the local community.

Effective marketing efforts will attract guests to a foodservice operation. When they arrive, or when they access a menu for call-in or online ordering, the specific menu items offered, and how they are described to potential guests, becomes extremely important.

In fact, the implementation of a well-thought-out menu that attracts guests and causes them to return is one of the most important factors in a foodservice operation's long-term success. The planning, design, and implementation of a foodservice operation's actual menu is so important that properly developing the menu will be the sole topic of the next chapter.

What Would You Do? 5.2

"I'm already spending a ton of time monitoring the sites and responding to the few negative reviews we receive," said Roz Jaffer, the assistant manager of Zaytoons' Mediterranean restaurant.

Zaytoons' was a popular operation featuring traditional Mediterranean dishes including Chicken Shawarma, Lamb Kabobs, Shish Kafta, and Kibbi. The previous manager of the restaurant had assigned Roz the task of monitoring UGC sites featuring customer reviews and responding to any negative reviews.

Now, Jessica Dora, the operation's newly assigned general manager, was asking Roz to also respond to some guests who had posted positive reviews.

"I know it's going to take some time," said Jessica, "but I think it's time well spent. If we only respond to our negative reviewers, then we're basically not valuing our happiest customers, only the angriest ones. I think our review scores are increasingly important. If it turns out that responding to some of the positive reviews takes you away from your other duties, then I think it's worth it to get you some help."

Assume you were Jessica. How would you justify the expense required to pursue this new course of action to the owners of your operation? Do you think the additional labor costs that may be incurred by providing Roz with additional help should be considered a marketing cost or a labor cost? Explain your answer.

Key Terms

Link	Customer loyalty	Up-selling
Local link	program	Marketing return on
Convention and Visitors	Paid social (advertising)	investment (ROI)
Bureau (CVB)	Arithmetic average	
Third-party delivery	Revenue metric	

Operator's 10-Point Tactics for Success Checklist

Evaluate your need for, and the current status of, each of the following operational tactics. For those tactics you think are important, but not yet in place, develop an action plan for its implementation including who will be responsible for the tactic's completion and the target date by which it should be completed.

				If Not Done	
Tactic	**Don't Agree (Not Done)**	**Agree (Done)**	**Agree (Not Done)**	**Who Is Responsible?**	**Target Completion Date**
1) Operator understands the importance of local links in the expansion of their online marketing presence.	—	—	—		
2) Operator has identified and established a relationship with the third-party-operated websites that are listed highest on the SERPs for their operation.	—	—	—		
3) Operator has carefully reviewed the potential advantages of partnering with third-party delivery apps when providing off-site delivery services for guests of their operation.	—	—	—		

Tactic	Don't Agree (Not Done)	Agree (Done)	Agree (Not Done)	If Not Done	
				Who Is Responsible?	Target Completion Date
4) Operator has carefully reviewed the disadvantages of partnering with third-party delivery apps when providing off-site delivery services for guests of their operation.	____	____	____		
5) Operator can state the reasons for the growing popularity and importance of all UGC sites.	____	____	____		
6) Operator has a system in place for regularly monitoring increases and decreases in the popularity of various UGC sites.	____	____	____		
7) Operator has identified UGC sites featuring user reviews that are commonly viewed by members of their target markets.	____	____	____		
8) Operator recognizes the steps that increase the number and quality of reviews posted on UGC sites featuring user reviews.	____	____	____		
9) Operator understands the importance of professionally written responses to negative reviews posted on popular UGC sites featuring user reviews.	____	____	____		
10) Operator recognizes the importance to success of regularly reviewing the overall effectiveness of the strategies and tactics used in their marketing efforts.	____	____	____		

6

Creating and Managing the Menu

What You Will Learn

1) The Importance of Effective Menu Design
2) How to Create a Food Menu
3) How to Create a Beverage Menu
4) How to Create Digital Display Menus
5) Special Concerns for Pickup Menus

Operator's Brief

In this chapter, you will learn that a menu is the list of items sold in a foodservice operation, and it is also the way information about available items is communicated to guests. An operation's menu is a powerful marketing and cost control tool, and it must be created professionally and in a way that communicates directly to an operation's target market (s).

There are different types of menus. Among the most important are the à la carte, table d'hôte, cyclical, and du jour menus. A foodservice operation's menu is best produced when all menu designers bring their unique perspectives to the menu's content, design, and development.

The development of an effective food menu is a four-step process: (i) identifying menu categories to be used, (ii) selecting individual menu items, (iii) writing menu copy, and (iv) designing the menu. The same four-step process is used to develop an operation's beverage menu(s).

Digital menus are an increasingly important marketing tool for all foodservice operations. Digital menus can be developed that are viewed on-site or on a user's computer or smart device. Not all foodservice operations require an on-site digital menu, but today all foodservice operations must develop menus suitable for viewing on guests' digital display devices. The reason: digital menus are increasingly the tool that current and potential guests use to make decisions about the foodservice operations from which they will (hopefully!) make purchases.

The increasing off-premise consumption of menu items requires operators to carefully consider which menu items should be offered on pickup (take-out) menus and how they are packaged and delivered to guests. Sometimes selected menu items may even be included or excluded from the pickup menu based primarily on their ability to be packaged and held with minimal loss of quality. In this chapter, you will learn how to identify menu items that are most suitable for off-site consumption.

CHAPTER OUTLINE

The Importance of the Menu
    The Menu as a Communications Tool
    Types of Menus
    The Menu Development Team
    Legal Aspects of Menu Design
    Creating the Food Menu
    Identifying Menu Categories
    Selecting Individual Menu Items
    Writing Menu Copy
    Key Factors in Successful Food Menu Design
    Beer Menus
    Wine Lists
    Spirits Menus
    On-site Digital Menus
    On-user Device Digital Menus
    Choosing Takeaway Menu Items
    Packaging Takeaway Menu Items
    Holding Takeaway Menu Items for Pickup or Delivery

The Importance of the Menu

The word **menu**, like many other foodservice terms, is of French origin. Literally translated, it means "detailed list." Foodservice operators use this definition when deciding the specific menu items to be available for guest selection and the prices they will charge for those items.

Key Term

Menu: A French term meaning "detailed list." In common usage it refers to (i) a foodservice operation's available food and beverage products and (ii) how these items are made known to guests.

The term "menu" has a second meaning when, for example, a server provides a copy of an operation's menu to a guest. Historically, this type of menu was most often made of paper, but today digital menus (addressed later in this chapter) are now becoming increasingly more popular. In this second context, "menu" refers to the ways available items are made known to guests. The focus of this chapter is primarily on this second usage of "menu" because the way foodservice operators communicate their menu items to guests is important to an operation's success.

The Menu as a Communications Tool

The menu, more than any other management tool, directly affects the marketing, financial, and operating success of every foodservice operation. An effective menu message, delivered through the proper marketing channel (see Chapter 3) will appeal to guests. However, unpopular menu items and those not prepared and presented to guests in a professional manner will not attract new nor retain existing customers.

All successful foodservice operators want to develop a menu that consistently offers quality products and services to their guests. Menus take on an important communication function when they deliver messages of quality food and beverage products. This is done with creative, well-thought-out, and accurately portrayed listings of the menu items to be served.

Key Term

À la carte (menu): A menu that lists and prices each menu item separately.

Key Term

Table d'hôte (menu): A menu with a pre-selected number of menu items offered for one fixed price. Pronounced "tah-buhlz-doht." Also known as a "prix fixe" (pronounced prē-'fēks) menu.

Types of Menus

When some foodservice operators consider their menus, they first think about meal periods (breakfast, lunch, and dinner) or the types of food and beverage items offered for sale (e.g. appetizer menus, dessert menus, or cocktail menus). Other operators may relate menus to where items are offered for sale (dining room menus, banquet menus, and takeaway menus).

Another good way to think about the menu is to consider how items are priced on it, and the frequency with which they are offered. The four most common examples of this approach are the **à la carte**, **table d'hôte**, **cyclical**, and **du jour menus**.

Key Term

Cyclical (menu): A menu in which items are offered on a repeating (cyclical) basis.

Key Term

Du jour (menu): A menu featuring items and prices that change daily. Pronounced "duh-zhoor."

À la Carte Menu

"À la carte" is a French term, and it refers to a meal in which guests select the individual menu items desired. Selected items are then prepared (or portioned) to order and are priced individually. For example, a guest in a casual-service restaurant that offers an à la carte dinner menu typically orders and pays separately for an appetizer, entrée (with salad and potato or vegetable), and dessert. Similarly, guests in a noncommercial cash cafeteria operation in a college or business dining operation may select their menu choices as they pass through the cafeteria's line and then pay for each selected item separately.

Table d'Hôte Menu

The French term, "table d'hôte," refers to a meal composed of menu items offered at a fixed price. In this case, the guest pays one price for all the items ordered. For example, a guest ordering dinner might receive an appetizer, entrée (with salad and potato or vegetable), and dessert and then pay a single fixed price for the entire meal. This would also be the case, for example, for guests in a noncommercial cash cafeteria operation such as a college or business dining operation who select a "dinner special" composed of an entrée, one or more accompaniments and, perhaps, dessert and a beverage and then pays a single fixed price. This type of menu is also known as a "prix fixe" (fixed price) menu.

Cyclical Menu

The term "cyclical" refers to the cycle or frequency with which a specific menu is repeated. Cyclical menus are most often used in noncommercial foodservice organizations where, for example, college students may be consuming all or most of their meals on-site. To reduce boredom and to reduce the need for menu planners to frequently plan new menus, a cyclical (cycle) menu might be used.

For example, in a senior citizen's assisted living facility a weekly menu may be created, and then that same menu would be offered to residents every six weeks. In this example, the foodservice operation is on a six-week cycle. Note: Cycle menus still offer operators the opportunity to vary menus to account for special occasions or holidays.

Du Jour Menu

The French term, "du jour" means "of the day." In other words, a du jour menu changes daily. Some operations, and especially fine-dining alternatives, may offer only a limited number of items daily. These change based on the quality of the ingredients that the proprietor or chef can purchase at the market that day. Numerous foodservice operations, however, may choose to offer "daily specials" such as a soup du jour, seafood du jour, or coffee du jour.

These four menu types are not mutually exclusive. Foodservice operations frequently offer combinations of more than one type. Consider, for example, the casual service restaurant with an à la carte menu that has a du jour (special-of-the-day) salad, soup, and entrée. Another example: the operation with a table d'hôte menu that changes daily (du jour). Similarly, a noncommercial facility such as a hospital can offer a cyclical menu but charge cash (on an à la carte basis) for hospital visitors dining in the cafeteria.

The Menu Development Team

Regardless of the type of menu being developed, a team of foodservice professionals most often plans the best menus. The reason: no single person generally has sufficient knowledge of everything needed to produce an outstanding menu.

For example, a chef or other member on the food production team can provide great knowledge about the ingredients contained in menu items and the methods used to best produce them. This information, of course, will be important as the menu is developed. Other foodservice professionals, however, will likely have expertise in how menus are designed, written, and produced. These marketing experts will be just as important for the development of an effective menu as are the operation's food production personnel.

An effective menu development team should include back-of-house food production staff, and it should include at least one member from the front-of-house staff as well. In fact, regardless of the formal titles of those who assist with menu planning, there should always be persons on the team who bring the following perspectives to the task:

1) The operation's owner or manager
2) The guests for whom the menu is written
3) The purchasers who procure the required items dictated by the menu
4) The food production staff who must produce items in the quantities and at the quality levels that are required
5) If applicable, beverage production staff who must produce the drinks to be served
6) The service personnel who will be responsible for the face-to-face delivery of menu items to guests
7) The menu designers; those responsible for writing, editing, proofing, layout, and menu format.

Of course, one individual could bring multiple perspectives to the task of menu planning. However, a foodservice operator must be careful not to assume that one person can bring all perspectives together in the menu planning process. When planning menus, it is generally agreed that the more collaboration involved in the development of the menu the better will be the final product.

Legal Aspects of Menu Design

Foodservice operators have much freedom as they write and design their menus, but that freedom is limited. Several Federal agencies place specific restrictions on what foodservice operators can state about their menu items. The best menus effectively communicate with their readers, and the accuracy of the information provided is an essential aspect of this effective communication. To comply with accuracy in menu laws, menu makers must truthfully represent the items being served.

Problems with menu accuracy can occur in several ways. Important areas to consider as menu makers create menus include references made about:

✓ *Quantity*. A two-egg omelet must contain two eggs; an eight-ounce steak should weigh a minimum of eight ounces (before cooking).

✓ *Quality*. The term "**prime**" when used to describe a steak refers to a specific U.S. Department of Agriculture (USDA) grading standard.

> **Key Term**
>
> **Prime (beef):** The highest quality grade given to beef graded by the USDA. Prime beef is produced from young, well-fed beef cattle, and it has abundant marbling (the amount of fat interspersed with lean meat).

U.S. Grade A or U.S. Fancy for vegetables and Grade AA for eggs and butter also indicate quality grades. Only the quality of the products actually used should be indicated on the menu. If it is not possible to ensure that the intended quality grade will always be available, it is best not to list the quality designation on the menu.

✓ *Price*. If there are extra charges (for example, for higher-cost liquors used to make drinks) these prices should be identified. If there are service charges (e.g. for groups larger than a specified size), these must be indicated.

✓ *Brand names*. If a specific product brand is specified (Coca Cola or Pepsi Cola, for example), this exact brand should be served.

✓ *Product identification*. For example, maple syrup and maple-flavored syrup are not the same. Neither are orange juice and orange drink or mayonnaise and salad dressing.

✓ *Points of origin*. "Pacific" shrimp cannot be from the Indian Ocean; Idaho potatoes are not from Wisconsin; and Lake Michigan whitefish should not be from Lake Erie.

✓ *Preservation methods.* For example, frozen apple juice should not be represented as "fresh," and canned green beans should likewise not be referred to as "fresh."

✓ *Food preparation.* "Made on-site" or "Chef made" products are different from a **convenience food** product made elsewhere; also, food should be prepared as stated on the menu. For example, a product described as "sautéed in butter" should not be sautéed in margarine.

Key Term

Convenience food: Any food item that is partially or fully prepared before it is purchased by a foodservice operation.

✓ *Verbal and visual presentation.* A menu photograph depicting, for example, eight shrimp on a shrimp platter should accurately represent the number of shrimp served; a specialty drink pictured in a fancy stemware glass with a unique garnish should be served in this same manner.

✓ *Dietary and nutritional claims.* The laws addressing dietary and nutritional claims on foodservice menus are perhaps the most extensive of all. This is reasonable because dietary claims are important for those who are health conscious, or those who may be allergic or very sensitive to certain foods. Among the menu terms that have been legally defined by the Food and Drug Administration (FDA) in the United States are:

- *Low fat.* The FDA has established that items referred to as "low fat" should have 3 grams of fat or less in a serving. However, if a restaurant serves a portion size larger than the FDA's standard portion, a "low-fat" food may contain a correspondingly larger amount of fat.
- *Lite.* This term can describe the taste, color, or texture of a food, or it may indicate that a food's calorie, fat, or sodium content is reduced. The menu must clearly indicate what "lite" means for the specific menu item.
- *Cholesterol-free.* Foods such as meat, poultry, and seafood naturally contain cholesterol. Menu planners should recognize that "cholesterol-free" does not mean the same as "fat free."
- *Sugar-free.* This term does not mean "calorie-free" or "fat-free." *Gluten-free.* This term can only be used when a menu item contains no gluten.
- *Healthy.* Foods noted to be "healthy" should be low in total fat and saturated fat, and they should not be high in cholesterol or sodium. There are not, however, limits on the amount of sugar or calories that a "healthy" food may contain.
- *Heart.* Menu claims such as "heart-healthy" or "heart-smart," and the use of heart symbols imply that a food may help reduce the risk of heart disease. Foods to which a heart-related claim has been made should be low in total fat, saturated fat, and cholesterol, and they should not be high in sodium.

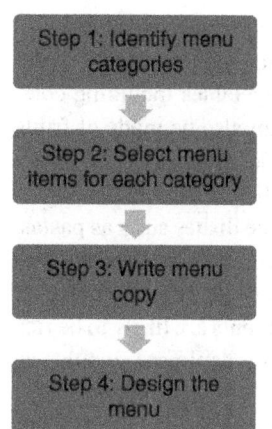

Step 1: Identify menu categories

Step 2: Select menu items for each category

Step 3: Write menu copy

Step 4: Design the menu

Figure 6.1 The Food Menu Development Process

After the menu development team has been selected and members have been instructed about legal requirements related to producing menu items, they can begin to create an operation's food menu.

Creating the Food Menu

As shown in Figure 6.1, the creation of a food menu is a four-step process:

Identifying Menu Categories

Most menu items are organized into logical categories on the menu. These groupings allow readers to consider choices among reasonably equivalent alternatives. Generally, foodservice operations with higher guest check averages (see Chapter 2) tend to offer more menu categories.

For example, a high-check average fine-dining operation may offer numerous appetizers, soups, salads, hot and cold entrées, vegetables, other accompaniments, desserts, and beverages along with a wine list. By contrast, a quick-service restaurant (QSR) may offer fewer categories (perhaps only sandwiches, side dishes, limited desserts, and beverages).

In many cases, a review of menus from similar operations and a careful analysis of the success of an operation's own menu can provide helpful information about the best categories to select when creating or revising a menu. Typically, however, most foodservice operators identify the following broad categories for their menus:

Entrées. When reading a menu many foodservice guests choose entrées first and then select other items that go with the selected entrée. Many guests think of entrées as hot items, and many are served hot. However, others such as a chef's salad may be cold. Still others may be hot and cold such as a cold Caesar salad topped with a hot grilled chicken or salmon fillet. Regardless of the items included in it, entrées may be the first menu items reviewed by guests so, for most operations, this is an essential menu category.

Appetizers. These items served before the meal to "tempt" one's appetite are often (but not always) smaller, bite-sized items. They may be hot (such as deep fried zucchini or crab dip) or cold (such as a shrimp cocktail or cheese and fruit). Note: Some light eaters may select one or more appetizers as their entree.

Soups. Some foodservice operations list soups as a separate category on the menu, and others may include them as part of the appetizer category or with salads.

Salads. Many operations offer two types of salads. Entrée salads (see above) and accompaniment (side) salads may be made from various types of lettuce and other greens. Salads may also be made from other vegetables including coleslaw (made from cabbage) and potato salad. Salads can also be made of fruits including oranges, melons, pears, apples, and pineapples.

Vegetables and other accompaniments. Vegetables including potatoes, asparagus, and green beans are offered on many menus. Other side dishes such as pastas, dumplings, and noodles are sometimes offered as a menu accompaniment. The specific dish offered is often dictated by the cuisine. For example, in Southwestern and Mexican operations the accompaniments are likely to be rice and beans. In an Italian restaurant, a pasta accompaniment would be more common.

Desserts. After-dinner sweets are often included on the menu with other items. Increasingly, however, a separate menu is used to feature these items. In many cases a dessert menu includes photos of the desserts being offered.

Beverages (nonalcoholic). Many foodservice operations offer guests a wide variety of nonalcoholic beverage choices. Traditional items such as coffees, teas, waters, and soft drinks may be listed on the menu. Sometimes, the brands of soft drinks (e.g. Coke or Pepsi) are also listed.

Foodservice operators must carefully select the categories to list on their menus. If too few categories are available, the menu may be confusing to guests. Selecting too many categories, however, can also be confusing. An effective menu lists the items available for sale and helps guests to easily make their selections.

Selecting Individual Menu Items

After menu planners have a basic structure in place (they have chosen their desired menu item categories), individual items to be included in each category are identified.

Typically, menu items in the entrée category are considered and selected first. Then many menu planners consider, in sequence:

✓ appetizers
✓ potatoes, rice, and other vegetable accompaniments
✓ soups
✓ salads
✓ desserts
✓ beverages

Important factors to consider when selecting individual menu items include:

✓ *Range (variety).* Normally there should be a range of temperatures, preparation methods, textures, shapes, and colors in the items that comprise a menu item category.

✓ *Temperature.* There is typically an expectation that some items will be served hot (meat, poultry, and seafood entrées), and other items will be served chilled (dinner and side salads, fruits, and cheeses).

✓ *Preparation method.* Meats can be grilled, fried, braised, or roasted. Vegetables can be boiled, baked, fried, or sautéed, and desserts can be served fresh, chilled, cooked, or baked.

✓ *Texture.* Alternative menu items can be soft, firm, or crunchy. They can be liquid or solid, dull or shiny, and wet or dry. Texture adds to the food's presentation, and it should be considered as items for each menu category are selected.

✓ *Flavor.* While menu planners have traditionally thought of flavor in terms of the basics (sweet, sour, salty, bitter, and **umami**), today's menu planners include additional flavors such as various degrees of hot (buffalo wings), spicy (Thai and Cajun dishes), and smoked (barbecue meats and poultry). While the concept of "taste" is very complex and involves more than just flavor, guests' perceptions about "what the food should taste like" is very important. When possible, menu makers should include a variety of flavors in each menu item category.

> **Key Term**
>
> **Umami:** "Essence of deliciousness" in Japanese, and its taste is often described as the meaty, savory deliciousness that deepens flavor.

✓ *Color.* Color is an integral part of a menu item's eye appeal. Color may be impacted by preparation method (example: the coating of fried chicken makes a brown product, and poaching the same chicken parts produces a white item). Good menu planners also recognize that the use of garnishes can help to assure that a menu item's color looks "right" on the plate.

Writing Menu Copy

Menu copy is the term foodservice operators use to indicate the individual descriptions of items listed on a menu. Well-designed menus effectively communicate with guests. As a result, the menu copy used for an item should address both what is being sold and answer basic questions that a menu reader might ask.

> **Key Term**
>
> **Menu copy:** The words and phrases used to name and describe items listed on a foodservice menu.

When producing menu copy menu designers should:

✓ *Write plainly.* The first responsibility of well-written menu copy is to describe menu items sufficiently so guests know what they will be buying. Unless the menu is for an ethnic-themed operation, foreign terms should be used sparingly. In addition to helping guests choose the items they want to buy, descriptive menu copy also reduces the order taker's time, and servers will not continually need to explain important details of menu items.

✓ *Tell guests what they should know.* An effective menu provides a sufficient description of an item so there will be few or no "surprises" when guests receive their orders.

✓ *Spell words correctly.* Menu makers must be careful to ensure the menu copy is grammatically correct and error free.

✓ *Write clearly, punctuate properly, and capitalize consistently.* While most menu readers do not expect menu writers to have perfect grammar, obvious errors on menus will likely be discovered (and pointed out!) by some guests. In addition, poor grammar may cause confusion. For example, in the following illustration, only the menu copy for item number 4 properly describes a sirloin steak "smothered" with onions and accompanied by a baked potato:

1) Sirloin Steak served with baked potato smothered in onions.
2) Sirloin Steak served with baked potato and smothered in onions.
3) Sirloin Steak smothered in onions and baked potato.
4) Sirloin Steak smothered in onions and served with baked potato.

Although descriptions 1 through 3 come close to accurately describing the item, the inadequate grammar and punctuation can make the items' confusing (or even comical!) to a menu reader.

The task of developing a good menu is just as difficult as (or more difficult than!) developing a good recipe. Significant time will be required as the menu evolves through the several (or more!) required development and writing phases. A final menu draft should be reviewed by the operation's owner, manager, production personnel, and others to help ensure it effectively explains necessary details in available menu items.

Key Factors in Successful Food Menu Design

The final step in the menu development process is its actual design. This is a complex step that often requires many important decisions. Frequently, menu designers place categories on the menu based on the order in which each is normally served. For example, appetizers will be listed before soups and salads, which are listed before entrées to be followed by desserts.

As menus are designed, some operators use sub-headings within menu item categories. For example, if an operation features several seafood, beef, chicken, and pork entrées, sub-headings to identify each of these entrée types can be helpful. When sub-headings are not used, menu designers may list items of a similar type (for example, beef items or poultry items) together in the entrée category rather than listing beef and poultry items randomly throughout the entrée list.

As menus are designed, the best operators recognize that menus have **"prime real estate"** areas in which the operation's most popular and highly profitable items should be located. Experienced operators understand that, over time, the profitability and popularity of menu items may change, and menu design changes may be required to accommodate these changes.

Key Term

"Prime real estate (menus"): A phrase used to define the areas on a menu that are most visible to guests, and which should contain the items menu planners most want to sell (those that are most popular and profitable)."

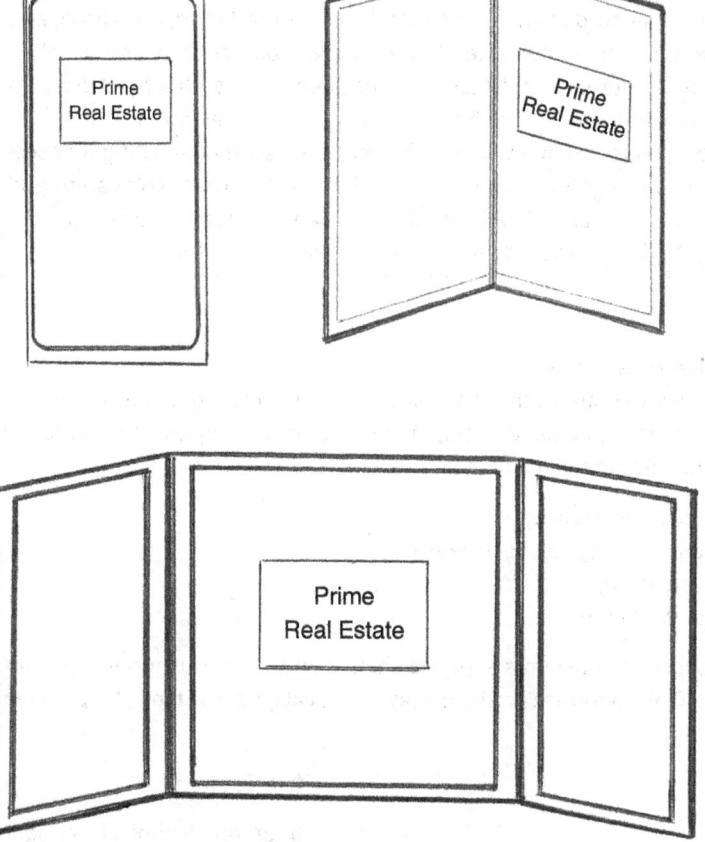

Figure 6.2 Prime Real Estate Areas on Various Menu Styles

Regardless of their final design choices, an operation's menu should always be easy to read, and it should not appear cluttered. Some experts even suggest that as much as 50% of the surface space on a menu should be left blank. While this may seem like a lot of wasted space, wide borders around the outside edges of the menu and between categories and the items within them help make menus easy to read.

Technology at Work

A growing body of research suggests there is a real science to effective menu design. Principles of psychology and buyer behavior are at work when menus are properly developed. To cite just one example, menu design professionals know that placing an expensive item at the top of a menu's category make items listed below that item look more reasonably priced.

An effective menu design should communicate the brand, the vision, and the appropriate guest experience offered by the foodservice operation. Most industry experts agree that effectively designed menus directly impact an operation's revenue producing ability and its guest satisfaction levels.

Fortunately, there are many companies that specialize in assisting foodservice operators in designing menus. To examine some of these companies and learn about the costs associated with this assistance, enter "Menu design specialists" in your favorite search engine and review the results.

Creating the Beverage Menu

Many foodservice operations that serve alcoholic beverages offer one or more separate menus for these items. Creating a beverage menu requires the same four steps used when designing the food menu:

Step 1: Select menu categories
Step 2: Select menu items for each category
Step 3: Write menu copy
Step 4: Design the menu

Just as foodservice operators separate their menus into categories, operators who serve alcoholic beverage products may create categories (Step 1) for the items they sell.

Key Term

Beer Menus

Technically, **beer** is a fermented alcoholic beverage made from malted grain and flavored with

Beer: An alcoholic beverage made from malted grain and flavored with hops.

hops. Lagers and ales are the two basic styles of beer, and some operators design their menus to separate beers based on these two basic styles. For example, an operator might create beer categories that result in listing lager beers on one part of the menu with ales on another part.

Stouts and Porters are additional beer types, and these may also be listed in a separate category.

Some operators create menus that categorize their beers based on packaging format rather than style. For these operators, beer categories include **draft beers**, canned beers, and bottled beers, and the operator creates a beer menu with items listed under these three categories.

> **Key Term**
>
> **Draft beer:** An unpasteurized beer product sold in kegs; also known as "tap" beer.

Yet another way for operators selling beer to create menu categories is by basing them on point of production. For these operators, beers are categorized as Domestic beers (those made in the United States) and Imported beers.

Due to their rising popularity, some foodservice operations create special category **craft beer** menus. These beers sell for a premium price and offer an effective way to optimize beer sales by using the beer menu to market locally made beers and microbreweries.

> **Key Term**
>
> **Craft beer:** A beer produced in limited quantities and with limited availability.

> **Key Term**
>
> **Wine:** An alcoholic beverage made from fruit; most typically grapes.

Wine Lists

Operators offering **wine** also make important decisions regarding the categories of products placed on their **wine lists**.

Many wine classification categories can be considered as operators produce their wine lists. These include:

> **Key Term**
>
> **Wine list:** A foodservice operation's wine menu.

1) Order of consumption

   When wines are categorized based on their order of consumption, the wine list is typically separated into Appetizer wines, Entree wines, and Dessert wines.
2) Country (and/or state) of origin

   When wines are categorized by their point of origin the wine list will have separate sections for each country whose wines are offered. For example, wines from Germany, France, the United States, and Italy would be listed separately. A point of origin wine list may also separate wines into the states from which they were produced. Examples: California, New York, Michigan, and Texas.

3) Color

Wine color is a popular method of wine list category creation. When categorizing by color an operation's wine list typically has categories of red, white, and rosé (blush).

4) Grape used for production

Wines can be made from nearly any type of grape, but those made from some types of grapes are more popular than others. When operators categorize their wine lists based on the grape used to produce the wine, product categories might include Chardonnay, Cabernet Sauvignon, Pinot Noir, Merlot, Zinfandel, and Syrah (among many others).

5) Serving format

Some operators produce their wine lists based on whether it is offered for sale by the glass, the **carafe**, or the bottle.

6) Carbonation

Some operators categorize wines based on whether they are "sparkling" (carbonated) or "still" (noncarbonated).

7) Selling price

When utilizing selling price to plan wine lists, wines are typically listed from most expensive to least expensive or from least to most expensive. It is best to remember that prices should be influenced by the type of operation selling them as well as the quality of wines sold.

Generally, each wine list should include some less expensive wines so cost-conscious guests can enjoy wine with their meals. Similarly, a wine list should include some higher quality and higher priced wines for those customers who enjoy drinking fine wines with their meals.

Operators who sell wine may categorize their selected product offerings using a combination of two or more classifications. For example, wines can be listed by country of origin and price.

> **Key Term**
>
> **Carafe:** A container used for serving wines. A standard carafe holds one standard size (750 ml) bottle of wine. However, carafe size may vary based on operators' service preferences.

Spirits Menus

Spirits are the most potent of alcohol beverages because they are distilled to increase their concentration of alcohol.

Often referred to as "Cocktail" menus, the development of a spirits menu is directly affected by the Alcohol and Tobacco Tax and Trade Bureau (TTB), a division of the U.S. Department of the Treasury. The TTB sets very specific requirements for the labeling of spirit products including:

> **Key Term**
>
> **Spirits (beverages):** Alcoholic beverages produced by the distillation of fermented grains, fruits, vegetables, and/or sugar.

✓ Brand or trade name
✓ Class: The broad category of "distilled spirits" is divided under standards of identity into general but defined classes, and examples include "Neutral Spirits or Alcohol" and "Whisky."
✓ Type: Under most general classes of spirits are specific, defined types of distilled spirits. For example, "Vodka" is a specific type of "Neutral Spirits or Alcohol," and "Straight Bourbon Whisky" is a specific type of "Whisky."
✓ Alcohol content in percent by volume (ABV). **Proof** is optional, but if stated it must be paired with the ABV.
✓ Net contents stated in metric units such as "750 ml" or 1.5 L."
✓ Name and address of the distiller for spirits produced in the United States or the importer of foreign spirits.
✓ Country of origin including where the product was made and bottled.
✓ A commodity statement which identifies the type of distillation utilized. This includes the grains used such as wheat or rye whiskey and whether the product was made using original distillation or redistillation.
✓ A health claims warning.

Key Term

Proof (alcoholic beverage): A measure of the alcohol content of an alcoholic beverage. In the United States, alcohol proof is defined as twice the percentage of "alcohol content in percent by volume (ABV)."

Foodservice operators may use any of the required labeling information on spirit bottles to create cocktail menus. Spirits are somewhat unique as a menu item because, in many cases, guests will not often "see" the products served before their consumption. That is, guests may be able to see that a served beverage "looks" like a "Gin and Tonic," but they may not see the actual type/brand of gin used to make the drink. This is why it is especially important that cocktails be produced using the exact ingredients and/or brands listed on the spirits menu.

It is also important to recognize there is a virtually unlimited number of product combinations that can be made as operators create spirit-based drinks. There is also an equally unlimited number of combinations of spirit products that can be made with various fruits, juices, flavored waters, and other ingredients. Therefore, operators serving spirit products have an almost unlimited number of choices related to the categories of their offerings of spirit products. Of most importance to the success of a foodservice operation is that spirit menus clearly indicate the spirit products used in the beverages produced.

Find Out More

The popularity of any specific spirit product can increase or decrease over time. Currently, whiskey (especially bourbon) consumption is rapidly increasing in the United States and worldwide. Note: "Whisky" refers to whiskies from Scotland, Japan, and Canada. "Whiskey" refers to the products made in the United States and Ireland.

Whiskey is a distilled spirit made from grains like corn and rye and is aged in wooden barrels. Bourbon is a type of whiskey, and there are strict rules in place to ensure its quality. Bourbon must be made in the United States, distilled from at least 51% corn, and be aged in new oak-charred barrels.

Contributing to the popularity of whiskey in the foodservice industry is the creation of numerous "Whiskey bars" or "Whiskey lounges." In these operations guests can sample a variety of different whiskies in a comfortable environment.

To learn more about the growing popularity of operations that feature whiskey as their primary menu item offering, enter "whiskey bar popularity" in your favorite search engine and view the results.

Technology at Work

The design and layout of spirit (cocktail) menus are among the most diverse of any foodservice menu. In addition to listing a drink's name and its price, spirit menus can include detailed descriptions of drinks, how they are made, and photographs of them.

In all cases an effective cocktail menu should encourage guests to drink responsibly and be consistent with the overall marketing message of the operation.

To see examples of the creativity that can be involved in developing a spirits menu, enter "Cocktail menu templates" in your favorite search engine and view the results.

What Would You Do? 6.1

"All I am saying is we should use one or two sentences to describe each wine we sell. That way our guests will know more about the exact type of wine they are buying and what entrees go best with the wines they choose," said Latisha, the wine steward at the Gaylord Bistro, a casual restaurant noted for its chef's imaginative entrée preparations. "It might also make their ordering faster."

"And what I am saying" replied Roland, the operation's dining room manger, "is that with the length of our beer menu, and the two-page specialty cocktail menu we already have, if we add even one or two sentences to describe each wine we sell the beverage menu is going to become massive."

Latisha and Roland were talking to Shivansh, the operation's general manager. All were members of the Gaylord Bistro's menu development planning team.

"Well," said Shivansh, "our current wine list includes about 50 different choices. I agree it's a good idea to tell guests as much as possible about what they're ordering, but my initial thought is that maybe this will actually slow down our dining room service as guests will spend a lot of extra time just reading our menus."

Assume you were Shivansh. After giving it some thought, would you likely agree with Latisha that detailed menus speed guest ordering or would you maintain your initial opinion that they will slow service? How can operators balance their guests' desire to know as much as possible about what they are ordering with the practical risk of potentially producing food or beverage menus that are too long and unwieldy to be effective?

Digital Display Menus

In the past, food and beverage menus were most typically printed on paper in a format that was intended to be easily read and used by guests. While these menus are still extremely prevalent, foodservice operators also increasingly create a **digital menu**.

A digital menu is a complex system of hardware and software to display dynamic menus on screens read by guests. In addition to the display of menu items and prices, digital menus can also display specials and promotions, pictures, videos, customer reviews, and more. Essentially, there are two main types of digital menus. The first type is intended to be displayed on-site, and another type is intended to be displayed on a user's computer or smart device.

Key Term

Digital menu: An integrated system that uses hardware and software to display an operation's menu on an electronic screen; also commonly referred to as a "digital display" menu or "digital menu board."

On-Site Digital Menus

Digital menus designed to be used on-site may be placed in an operation's interior to enable guests to approach, for example, a QSR operation's order, and then view a digital menu before placing orders. These on-site digital menu boards are typically colorful and include high resolution images of the menu items being served. These menu boards can be easily changed during the day to display, for example, an operation's separate breakfast, lunch, and dinner menus.

Digital menus may also be designed to be viewed from an operation's exterior and are often seen in foodservice operations with takeaway (drive-thru) service. When placed outside, they must be weatherproof to shield the hardware from environmental elements. They must also be properly designed to be easily read in bright sunlight and at night and constructed to minimize damage from vandals or careless drivers.

The hardware in a digital menu board consists of display screens and a media player: the device that downloads menu content from the sign's software. The **content management system (CMS)** in a digital menu board allows operators to create the content guests will view by uploading new content and then changing or modifying the content as needed.

Key Term

Content management system (CMS): Computer software used to load content into a digital menu display system.

Technology at Work

A well-designed content management system (CMS) allows foodservice operators great flexibility in using their digital menus.

Some CMS software makes it easy for operators to schedule menus by time of day so specific menu items appear only during a defined time. They also make it easy to load videos and animations to time and date-specific promotions on new products and higher margin impulse items.

In many cases, multi-unit operators can select a CMS that allows them to create location-specific content. Then, for example, an operation located in a suburban area could, at a given time of day, promote different menu items than a similar operation located in an urban area.

Improvements in CMSs are occurring rapidly and continually. To see some features in systems currently available, enter "content management systems for restaurant menus" in your favorite search engine and review the results.

On-User Device Digital Menus

Viewing menus and ordering online is not new, and pizza restaurants were pioneers in online ordering. For example, Pizza Hut first began experimenting with online ordering in 1994, and it launched its first mobile ordering app in 2009.

Today, foodservice guests' use of digital devices to order menu items is increasing steadily. According to a 2022 survey of users, digital ordering represents 28% of all restaurant orders.[1] Rather than view a paper menu, however, these guests

---

1 https://madmobile.com/blog/online-ordering-stats/ retrieved August 29, 2023

are viewing an operation's menu on their own user device (typically a cell phone or other handheld smart device).

There are significant differences between viewing a paper menu and a menu online. Operators must address these differences when posting menus online, and three of these important differences are page size, font size, and file size.

Page Size

Operators must recognize that a standard 8.5″ × 11″ piece of paper is about three times as wide as the average (4 inches) smartphone screen, and printed menus might even be larger than a standard size piece of paper. Unfortunately, operators who simply take a photo of their printed menus for online posting make it difficult for potential guests to read and navigate these online menus.

The better approach is to post menu images online that will automatically scale to fit the width of a mobile phone screen. Unlike print menus, there are no page boundaries or height limitations involved in designing a digital menu for viewing on a smart device. Therefore, operators can stack their content into one column and allow the volume of content to determine the page height. A smart device's screen automatically limits the content seen at one time. This enables users to scroll through the menu at a comfortable pace in much the same way as when they check their e-mails or read a news article online.

Font Size

A second important factor to remember is that the **font size** best used for paper menus differ from those best used for viewing menus on a user-owned smart device.

Font size is measured in points (pt.) that dictate lettering height. There are approximately 72 (72.27) points in 1 inch (2.54 cm). As a result, for example, a font size letter of 72 pt. would be about 1 inch tall, and a font size letter of 36 pt. would be about one half of an inch tall.

Key Term

Font size: The size of the characters printed on a page or displayed on a screen. Commonly referred to as "type size."

Figure 6.3 shows how various font sizes would appear on a typical user's 4-inch smart device screen.

Some general font sizing recommendations based on a smart device page width of 4 inches are:

✓ Category Headings (e.g. Appetizers, Salads, Entrees, and Desserts) 28–32 pt.
✓ Menu item names: 14–20 pt.
✓ Menu item descriptions: 12–14 pt.

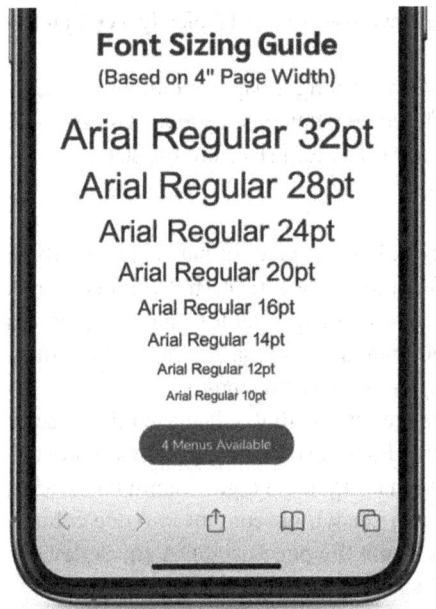

Figure 6.3 Font Sizing Examples for Smart Devices

As with printed menus, type sizes that are too small to easily read on a user's screen can be irritating and create serious mistakes when guests order. Most importantly, if guests cannot easily read a menu and are continually forced to manually expand their screens to do so, they may simply decide to order from another operation.

File Size

Smart device users download content at various speeds depending on their internet or phone connections and the quality of device being used. It is best for operators to keep menu files as small as possible to optimize them for web viewing. In general, the smaller the file the faster it will load on a smart device. Therefore, operators should have separate menu files for each of their major menu categories. Then, for example, a guest interested in ordering an entree need not wait for dessert content to be loaded before proceeding with their order.

Operators should understand that videos, detailed images (high resolution photos), and complex graphics displayed on a screen require more data download time than text downloads. If they want to show a variety of images such as dining room interiors or menu items, they should select such images carefully and only use them when they significantly add to the viewer's menu reading experience.

Technology at Work

Digital and QR code scan menus viewed on a smart device differ from print menus. When creating a printed menu, the amount of information presented should be minimized to reduce the menu's length and the number of required print pages. Since digital menus are not printed, their length is not typically of significant concern. Operators may continually add information since viewers know they can easily scroll up and down on web displays to view content.

Alternatively, high resolution photos ideal for use on a print menu may frustrate smart device users because of the download time and data usage involved in downloading and viewing them.

Many operators will likely require assistance to properly convert their print menus to digital menus suitable for smart device viewing, customer item selection, and mobile payment. Fortunately, numerous companies specialize in assisting operators in designing menus for smart devices. To examine some of these companies and learn about the costs associated with this assistance, enter "Converting print menus to mobile friendly menus" in your favorite search engine and review the results.

Special Concerns for Off-Premise Menus

An **off-premise menu** is one used by guests who wish to purchase menu items that will be picked up or delivered to them. An off-premise menu can be identical to an operator's normal menu, but in most cases it should not be. The reasons are many, but all relate to the challenges in maintaining product quality for off-premise menu item consumption. While not every foodservice operation will want to produce a separate off-premise menu, there can be significant advantages to doing so.

Key Term

Off-premise menu: A menu listing items available for pickup or delivery; also commonly referred to as a takeaway, take-out, or carry-out menu.

The COVID-19 pandemic of 2020–2022 impacted foodservice operations in many important ways. Some operations that never offered menu items for takeout or delivery had to do so to stay in business. Also, some operations that had previously generated only a relatively small percentage of their revenue from off-premise sales found that these sales represented most or even all of their entire revenue. What these operators discovered was that there are special concerns related to selecting, packaging, and holding menu items to be consumed off-premise.

Choosing Takeaway Menu Items

Menu items to be sold for off-premise consumption must be carefully chosen because the time between order pickup and item consumption is not known in advance. For guests picking up menu items and transporting them to a nearby location the time between pickup and consumption may be very short. Alternatively, on a busy night when a guest order has been prepared for pickup and delivery by a third-party delivery service, the length of time items will be held before consumption may be relatively long.

Menu item selection for takeaway menu items is important because guests always equate an operation's quality standards with the products they receive when they are picked up or delivered. Some menu items can be held for a relatively long time with no reduction in quality. Examples include pizza, fried chicken, deli sandwiches, and undressed salads. Some menu items, however, do not travel well. Examples are many deep-fried items (including French fries!), ice cream-based items, sauced items such as Eggs Benedict, souffles, and delicate fish dishes.

Each foodservice operation is different. Operators should carefully consider whether items that deteriorate if not quickly consumed should be offered on their off-premise menus. Troublesome items that do not reflect the quality of the entire menu can be omitted in favor of items that can be transported while preserving food quality, texture, and flavor.

Packaging Takeaway Menu Items

The packaging of takeaway items is just as important as their selection for inclusion on a takeaway menu. For example, assume an operator has just received an order for a green salad, hamburger, French fries, and a cold fountain drink. The order will be picked up and delivered by a third-party delivery partner to a customer who lives several miles away. If this guest's order is not packaged correctly, the salad may be wilted (if dressing was added before the salad's packaging), and the hamburger and fries may be cold (if an appropriate hot foods container was not utilized). As well, the cold fountain drink may contain melted ice and could be spilled during delivery if an appropriate cold food container was not utilized, and its lid did not fit snugly.

In this example guest satisfaction will be negatively impacted not because of the products selected, but because the selected products were improperly packaged.

As a second example, assume a foodservice operation received a pickup order for a three fish taco meal. The fish tacos *could* be fully assembled and placed in a Styrofoam container, and the food would technically be ready for pickup. It would be much better, however, if (a) fish fillings were packaged in a hot container, (b) items such as sour cream, cheese, and shredded lettuce were packaged in a cold container, and (c) warm tortillas were wrapped in tin foil. In this second packaging example, even if the guest's pickup time were delayed, product quality would be optimized.

While each foodservice operation is different, fundamental principles of successful takeaway food packaging include:

✓ Packaging hot foods separate from cold foods
✓ Packaging liquid foods in containers with secure fitting lids
✓ Utilizing see-through containers whenever possible
✓ Replacing plastic containers with more environmentally friendly containers, when possible
✓ Customize takeaway containers and products with the operation's name or logo when it is economically feasible to do so

---

Find Out More

Takeaway foods must be packaged properly for food safety and product quality. While many foodservice operations utilize Styrofoam clamshells as their go-to packaging option and Styrofoam does help keep hot foods hot and cold foods cold. However, these containers can break open easily and lead to messy leaks at the slightest bit of turbulence. Therefore, many operators increasingly look to other packaging options that provide a strong seal to keep all ingredients intact. In addition, most foodservice operators include condiments, appropriate cutlery, and napkins with their takeaway orders.

Fortunately, foodservice operators can now choose from a number of alternative products that help maintain the quality of takeaway foods and that are environmentally friendly. Examples include bamboo plates, wood dinnerware, and items made from sugar cane or bagasse (a by-product of sugar cane fiber) dinnerware. Other items include kraft paperboard made from 100% unbleached paperboard with an additional coating that makes the product cut-and leak-resistant.

To learn more about these and other new packaging options enter "Innovative takeaway food packaging" in your favorite search engine and view the results.

Holding Takeaway Menu Items for Pickup or Delivery

Even when menu items are carefully selected for suitability as takeout items are packaged properly, product quality can still decline. This is especially so if these items are not held properly until they are picked up by guests or delivery drivers.

Guest pickup of ordered items can be delayed by traffic problems, other unforeseen circumstances causing delays, or simply by miscalculation of travel times. Order pickup by third-party drivers can be affected by issues including the total number of deliveries requested at a specific time (some periods are busier than others). As well, the number of available drivers (with third-party delivery), and even the amount guests have indicated they intend to tip the driver are examples of other potential issues.

Regardless of the reasons, however, some takeaway orders will be held on-site for extended periods of time, and these longer holding times can result in product quality deterioration. To minimize the negative impact on food quality of excessive holding times, operators must ensure that hot food packages are held separately from cold food packages. In many cases, this will require separate containers or bags for a single pickup order. If hot and cold foods must be packaged together in the same bag, box, or container, the order should be assembled upon a guest's or a driver's arrival when possible.

Holding areas for takeaway items should be clean and brightly lit for ease in labeling and packaging orders. For guest orders that will be delivered in coolers or insulated bags, these packing items should be properly cleaned, sanitized, and disinfected between uses. Increasingly, some operators add tamper-evident labels (those that, when ripped or torn, alert guests that their order was tampered with), or tamperproof boxes can also be used for their deliveries.

The menu is a vital component of every foodservice operation. However, it is also essential that foodservice operators price the menu items they sell in a way that allows them to cover their operating costs and still provide an optimal profit level. The ability to properly price menu items is vitally important to a foodservice operator's success, and it will be the sole topic of the next chapter.

Technology at Work

As takeaway meals and third-party delivery companies continue to grow in popularity, the need for appropriate packaging continues to evolve.

Peace of mind for customers receiving delivered meals is important, and this increases the demand for tamperproof or tamper-evident packaging of delivered meals. This type of packaging helps ensure that food has not been "sampled" or otherwise adulterated by delivery drivers. Tamperproof packaging provides peace of mind to guests and foodservice operators, and guests can be assured delivered menu items are safe and unadulterated. Foodservice operators can then use third-party delivery services with reduced fear of the tarnishing of their operations' names by unscrupulous delivery drivers.

Innovative packaging for takeout and delivery foods continues to evolve. To learn more about tamperproof products resulting from this evolution, enter "Tamperproof packaging for restaurant delivery" in your favorite search engine and review the results.

What Would You Do? 6.2

"I can't believe this guest!" exclaimed Mohith as he looked over the shoulder of Rose, the owner of Smiley's Pizza (a ghost kitchen that serves take-out pizza and Italian-style appetizers and desserts). Mohith, the operation's kitchen manager, and Rose were looking at Rose's smartphone and reading a guest's online review of the operation. The review had just recently been posted on a popular social media site.

"This guest says that their pizza was fine, but the Fried Mozzarella Sticks were cold and soggy when they were delivered," said Rose.

"I was working in the kitchen when the guest placed the order. We were really busy, and I can't remember the exact order, but I do know we didn't have any problems getting all orders out on time," said Mohith.

"Then I guess we need to look at a potential delivery issue," said Rose. "If our third-party delivery partner picked up more than one order from us at the same time that night and, if this guest's order were delivered last, that could explain the problem. Regardless, the bottom line is that this guest only gave us a 1-star rating!"

Assume you were Rose. Do you think the guest cares whose fault it was that their mozzarella sticks were cold and soggy? Do you think this guest's complaint was primarily due to a production failure, a service failure, or a take-out menu design failure? Why do you think so and what should your team do now?

Key Terms

Menu
Á la carte (menu)
Table d'hôte (menu)
Cyclical (menu)
Du jour (menu)
Prime (beef)
Convenience food
Umami
Menu copy

Prime real estate (menu)
Beer
Draft beer
Craft beer
Wine
Wine list
Carafe
Spirits (beverages)

Proof (alcoholic
   beverage)
Digital menu
Content management
   system (CMS)
Font size
Off-premise menu

Operator's 10-Point Tactics for Success Checklist

Evaluate your need for, and the current status of, each of the following operational tactics. For those tactics you think are important, but not yet in place, develop an action plan for its implementation including who will be responsible for the tactic's completion and the target date by which it should be completed.

Tactic	Don't Agree (Not Done)	Agree (Done)	Agree (Not Done)	If Not Done	
				Who Is Responsible?	Target Completion Date
1) Operator understands the importance of utilizing the menu as a communications tool.	____	____	____		
2) Operator has carefully selected a menu development team to assist in the creation of the operation's needed menus.	____	____	____		
3) Operator has reviewed the basic requirements of truth-in-menu laws and how they affect menu development.	____	____	____		
4) Operator recognizes that initial steps in creating a menu are identifying menu categories and selecting individual menu items.	____	____	____		
5) Operator recognizes the final steps in menu development are writing menu copy and properly designing the menu.	____	____	____		
6) Operator understands the basic alternatives available when creating beer menus.	____	____	____		
7) Operator understands the basic alternatives available when creating wine lists.	____	____	____		

Tactic	Don't Agree (Not Done)	Agree (Done)	Agree (Not Done)	If Not Done	
				Who Is Responsible?	Target Completion Date
8) Operator understands the basic alternatives available when creating a spirit menu.	___	___	___		
9) Operator has reviewed and recognizes the challenges related to digital menu development.	___	___	___		
10) Operator has reviewed and selected off-premise menu items based, at least in part, on each item's ability to be packaged, held, and delivered properly.	___	___	___		

7

Successful Menu Pricing

What You Will Learn

1) The Relationship Between Foodservice Prices and Profits
2) The Factors Affecting Foodservice Prices
3) The Methods Used to Price Menu Items
4) How to Evaluate Pricing Efforts in Foodservice Operations

Operator's Brief

In this chapter, you will learn that properly pricing menu items is an important skill for all foodservice operators even though prices are viewed differently by sellers and buyers. For operators, costs and profits are critical when establishing menu prices, but guests are concerned about the value they receive. Therefore, in addition to pricing menu items to be profitable, you must also ensure menu prices communicate real value to your guests.

Numerous factors influence the prices you will charge for menu items. Among the most important are economic conditions, competition, level of service, type of guest, and product quality and costs. Additional pricing factors include portion size, delivery style, meal period, location, and bundling: a pricing strategy that combines two or more menu items that are then sold at a price lower than that of the bundled items when purchased separately. In this chapter, you will learn about these important factors.

When establishing menu prices some operators use a product cost-based approach. They believe a menu item's production cost relative to its price is of most concern. When using this pricing approach, menu items with lower product cost ratios are thought to be more desirable to sell than those with higher product cost ratios. Other operators use a more profit-oriented approach to establish their menu prices. Menu items providing the highest profit-per-sale are considered more desirable to sell than items with lower profit levels.

Regardless of the pricing approach used, it is important to regularly evaluate menu items that are the most popular and profitable. Then, you can modify or eliminate poor selling or unprofitable items and better promote those with the most popularity and profitability. In this chapter, you will learn how to successfully complete these essential tasks.

CHAPTER OUTLINE

Pricing Menu Items
    The Importance of Price
    The Operator's View of Price
    The Guest's View of Price
Factors Affecting Menu Pricing
    Economic Conditions
    Local Competition
    Level of Service
    Type of Guest
    Product Quality
    Unique Alternatives
    Portion Size
    Meal Period
    Location
    Bundling
Methods of Food and Beverage Pricing
    Cost-based Pricing
    Contribution Margin-based Pricing
Evaluation of Pricing Efforts
    Calculating Menu Item Popularity (Number Sold)
    Menu Modifications

Pricing Menu Items

Foodservice operators must successfully price their menu items to help ensure the long-term profitability of their businesses. Experienced foodservice operators know that, if menu prices are too low, an operation may be popular but not profitable. Alternatively, if menu prices are too high, the popularity of a foodservice operation will likely suffer because few guests will regularly visit. Operators who establish menu item prices must understand how to do so effectively, and this chapter addresses this topic.

The Importance of Price

Price plays a large role in a foodservice operation's profitability. To best understand the importance of price, foodservice operators must understand that the term as used in the foodservice industry has two separate definitions.

Note that, in both uses (noun or verb), the concept of an *exchange* between a buyer and a seller is important. The foodservice operator gives up (exchanges) a menu item when it is purchased, and the foodservice guest gives up (exchanges) the item's selling price for the menu item.

Foodservice operators can typically charge any prices they want to charge. However, potential guests can accept or reject the operator's opinion that prices charged are fair and will provide good value to them (the guests). In fact, in many cases, a foodservice operation will be selected by a guest based primarily on prices charged and the guest's perceptions of those prices.

Key Term

Price (Noun): A measure of the value given up (exchanged) by a buyer and a seller in a business transaction.

For example: "The price of the Mushroom Swiss Burger combo meal is $9.95."

Price (Verb): To establish the value to be given up (exchanged) by a buyer and a seller in a business transaction.

For example: "We need to price the Mushroom Swiss Burger combo meal."

To best understand pricing in the foodservice industry, it is important to first recognize that price is viewed very differently from the perspectives of an operator (the seller) and a guest (the buyer).

The Operator's View of Price

In most cases, foodservice operators can determine the products to sell, where and when they will sell them, and how they communicate selling prices to potential guests. Foodservice operators can propose their prices, but they also face the possibility that potential guests may *not* support their **value proposition**.

Price and value are closely related concepts in the foodservice industry. It is often stated that

Key Term

Value proposition: A statement that clearly identifies the benefits an operation's products and services will deliver to its guests.

the value of any item is equal to what a buyer will pay for it. If this is true, when a sale is *not* made, the buyer either believed the item was not worth the asking price or a lower cost alternative was available that was also considered to be worth its asking price.

From the perspective of most operators, a fair selling price would be an amount equal to the operator's incurred costs plus a reasonable (desired) profit.

Stated mathematically that concept is:

Item Cost + Desired Profit = Selling Price

When calculating menu item costs, operators must consider the estimated costs of food, beverage, labor, and all other expenses required to operate their businesses. Note: In most cases, professional foodservice operators can calculate these costs quite accurately.

The question of *"What is a reasonable desired profit?"* is more subjective. An operation's **profit margin** is the percentage of each buyer's revenue dollar the operation retains as profit. The higher the profit margin the more profitable the operation.

Key Term

Profit margin: The amount by which revenue in a foodservice operation exceeds its operating costs.

In most cases a foodservice operator wants to generate a reasonable profit margin because this is necessary to stay in business and to receive a fair return for the investment risks in the business. Therefore, consideration of operating costs incurred and the profit desired are the two most important concerns as operators establish menu prices.

The Guest's View of Price

All guests desire good (or better!) value for the products and services they purchase. If they do not believe good value was received, the guests are unlikely to be satisfied and probably will not return to the business. Foodservice operators must remember that, in all business transactions, value (see Chapter 1) is determined by the *buyer*, not the seller.

For example, a foodservice operator may sincerely believe that an item's quality, quantity (portion size), and delivery method will provide good value to guests if it is sold for $19.95. If, however, too few customers share that view of value, the menu item will not be frequently sold. Stated mathematically, a buyer's perception of value is expressed as:

Buyer's Perceived Benefit(s) − Price = Value

When making a purchase, buyers want to receive *more* value than the value of what they are giving up. Stated another way, there are three possible buyer reactions to any seller's proposed selling prices. These are shown in Figure 7.1.

When buyers believe the benefit they will receive is *less than* zero (less than what they give up in the exchange), they generally will not buy. When the benefit is *greater than* zero, guests are very likely to buy. When the perceived benefit is *equal to* "0," buyers are often indifferent to the purchase. In this scenario, if no

Buyer Assessment	Purchase Decision
1. Perceived Benefit – Price = A value less than "0"	Do not buy the item
2. Perceived Benefit – Price = A value equal to "0"	Do not buy in most cases
3. Perceived Benefit – Price = A value greater than "0"	Buy the item

Figure 7.1 Buyer's Assessment of a Seller's Value Proposition

other alternatives are available, they may make a purchase decision. If alternatives are available, buyers will likely consider these alternatives before making purchase decisions.

To best understand how buyers' perceived benefit assessments directly impact purchasing decisions, operators must first understand the concept of **consumer rationality**.

Consumer rationality assumes customers consistently exhibit reasonable and purposeful behavior, and buyers' generally make purchase decisions based on the belief that they (the buyers) benefit from doing so.

Key Term

Consumer rationality: The tendency to make buying decisions based on the belief that the decisions will result in a personal benefit.

For foodservice operators, the acceptance of consumer rationality involves a willingness to look beyond the obvious and then attempt to understand exactly how buyers believe they benefit from a business transaction. In some cases, this is not easy.

First, foodservice operators must resist the temptation to declare their guests are irrational (e.g. when operators criticize guests who state the operator's prices are "too high!").

All buyers like low prices, but what they seek most is value. Most foodservice guests are indifferent to the actual costs of operating a foodservice business and, therefore, they are indifferent to an operator's profit margin.

Also, rational buyers *do not* automatically equate a seller's proposed price with the amount of value they (buyers) receive in an exchange. In fact, conventional wisdom advises them not to do so. From common sense and even from a legal perspective, buyers assessing a seller's value proposition are cautioned not to trust the seller. As a result, *Caveat Emptor,* the Latin phrase for "let the buyer beware," is known and well-understood by most consumers.

Since buyers can be skeptical about a seller's initial value proposition, the foodservice operator's responsibility when pricing menu items is to ensure guests understand the answers to questions about *"What do I get?"* and *"Why is it of value?"* just as much as they understand the actual prices they will pay. Only then can buyers, who are increasingly sophisticated and web-savvy consumers, learn they will consistently receive *more* than the value of the money they pay when making their purchase decisions.

The foodservice operator's primary motivation (to recover operating costs and generate a profit) is very different from their guests' motivation to optimize the value received for the prices they pay. Therefore, while operators are wise to be concerned about their costs and profits, what they should be *more* concerned about is using the selling price as a means of communicating excellent value to their guests.

Factors Affecting Menu Pricing

When foodservice operators find that profits are too low, they frequently question whether their prices (and therefore their revenues) are too low. But it is important to recognize that the terms "revenue" and "price" are not the same thing. "Revenue" is the amount of money spent by all guests, and "price' refers to the amount charged for one menu item. Total revenue for a menu item is generated by the following formula:

$$\text{Selling Price} \times \text{Total Number Sold} = \text{Total Revenue}$$

There are two components of total revenue. Selling price is one component, and the other is the total number of items sold. Note: generally (but not always!), the economic **law of demand** indicates that, as an item's selling price increases, the number of that item sold decreases. Also, when an item's selling price decreases the number of that item sold will most often increase.

Experienced foodservice managers know that increasing prices without giving added value to guests result in higher selling prices but (frequently) lower total revenue because there is a reduction in the number of guest purchases that are made. For this reason, those factors affecting menu prices must be carefully evaluated because they can directly impact the prices operators are able to charge for the menu items they sell. Among the most important variables to analyze are:

Key Term

Law of demand: The law of demand holds that the demand level (number of units sold) for a product or service declines as its price rises and increases as the price declines.

✓ Economic conditions
✓ Local competition
✓ Level of service
✓ Type of guest
✓ Product quality
✓ Unique alternatives

✓ Portion size
✓ Delivery style
✓ Meal period
✓ Location
✓ Bundling

Economic Conditions

The economic conditions existing in a local area or even the entire country can significantly impact the prices operators charge for menu items. For example, a robust and growing local economy generally enable foodservice operators to charge higher prices for items they sell. In contrast, a local economy in recession and/or weakened by other events can limit an operator's ability to raise or even maintain current prices in response to rising product costs.

In most cases, foodservice operators cannot directly influence the strength of their local economies. They can and should, however, monitor local economic conditions and carefully consider the impact of these conditions when establishing menu prices.

Local Competition

The prices charged by an operation's competitors can be important, but this factor is sometimes too closely monitored by inexperienced foodservice operators. It may seem to some operators that their average guest is only concerned with low prices and nothing more. However, small price variations generally make little difference in the buying behavior of the average guest.

For example, if a group of young professionals goes out for pizza and beer after work, the major determinant is not likely whether the selling price for a small pizza is $17.95 in one operation or $19.95 in another. Other factors such as perceived product quality, location, and parking availability, for example, may be more important to these buyers than price.

The selling prices of potential competitors are of concern when establishing a selling price, but experienced operators understand that a specific operation can always sell a product of lower quality for a lesser price. While competitors' prices can help an operator arrive at their own property's selling prices, it should not be the only determining factor.

The most successful foodservice operators focus on building guest value in their operations and not in attempting to mimic competitors' efforts. Even though operators may believe their guests only want low prices, it is important to remember that consumers often associate higher prices with higher quality products and, therefore, products that provide a *better* price/value relationship.

Level of Service

The service levels an operation provides directly affects the prices the operation can charge. Most guests expect to pay more for the same product when service levels are higher. For example, a can of soda sold from a vending machine is generally less expensive than a similar- sized soda served by a service staff member in a sit-down restaurant.

Service levels can impact pricing directly and, as the personal level of service increases, selling prices may also increase. Personal service ranges from the delivery of products to a guest's home to the decision to quicken service by increasing the number of servers in a busy dining room (which improves service quality by reducing the number of guests each server must assist).

These examples should not imply that extra income from increased menu prices is necessary only to pay for extra labor required to increase service levels. Guests are willing to pay more for increased service levels. However, higher prices must cover extra labor costs and provide extra profit as well. The best foodservice operators survive and thrive over the years because of an uncompromising commitment to high levels of guest service, and they can charge menu prices reflecting enhanced service levels.

Type of Guest

All guests want good value for the money they spend. However, some guests are less price sensitive than others and, as a result, the definition of what represents good value often varies based on the type of clientele served. Consider the pricing and purchasing decisions of convenience store customers across the United States. In these facilities, food products such as pre-made sandwiches, fruit, drinks, cookies, and other items are often sold at relatively high prices. The customers in these stores most desire speed and convenience, and they are willing to pay premium prices for their purchases.

Similarly, guests at a fine-dining steakhouse restaurant are less likely to respond negatively to small variations in drink prices than are guests at a neighborhood tavern. A thorough understanding of exactly who the potential guests are and what they value most is critical to the on-going success of foodservice operators as their menu prices are established.

Product Quality

A guest's quality perception of a menu item can range from very low to very high, and perceptions are most often the direct result of how guests view an operation's menu offerings. These perceptions are directly affected by a menu item's quality, but they should never be shaped by the guest's view of an item's wholesomeness or safety. All foods offered by a foodservice operation must always be wholesome and safe to eat.

Guests' perceptions of quality are based on numerous factors of which only one is the quality of actual raw ingredients. Visual presentation, stated or implied ingredient quality, portion size, and service level are additional factors that impact a guest's view of overall product quality.

To illustrate, consider that, when most foodservice guests think of a "hamburger," they think of a range of products. A "hamburger" may include a rather small burger patty placed on a regular bun, wrapped in waxed paper, served in a sack, and delivered through a drive-through window. If so, guests' expectations of this hamburger's selling price will likely be low.

If, however, the guests' think about an 8-ounce "**Wagyu beef**-burger" with avocado slices and alfalfa sprouts on a fresh-baked, toasted, whole-grain bun and served for lunch in a white-tablecloth restaurant, the purchase price expectations of the guests will be much higher.

A foodservice operator can select from a variety of quality levels and delivery methods when developing product specifications, and as they plan their menus and establish selling prices. The decisions they make will have a direct impact on menu pricing.

Key Term

Wagyu beef: Beef from a Japanese breed of cattle that is highly prized for its marbling and flavor. In the Japanese language, "Wa" means Japanese, and "gyu" means cow.

For example, if a bar operator selects an inexpensive Bourbon to make their whiskey drinks, they will likely charge less for whiskey drinks than another operator selecting a better (higher quality and more expensive) brand. Guest perceptions of the value received from those buying the lower-cost whiskey drinks will likely be lesser than guests served a higher quality product.

To be successful, foodservice operators should select the product quality levels that best represent the anticipated desires of their target market (see Chapter 3) and their operations' own pricing and profit goals.

Unique Alternatives

Property operators must recognize that the overall experience of guests is part of the value received and menu pricing decision. For example, most guests agree that they would pay more for the views presented in an elegant dining room. Think about an ocean view and even lights that reflect these scenes after dark. Perhaps there is a local entertainer who performs regularly at the property, and/or property amenities including artwork and museum pieces. The total "value equation" considered by guests can often include features in addition to the food and beverage products being purchased.

Portion Size

Portion size often plays a large role in determining a menu item's price. Great chefs are fond of saying that people *"Eat with their eyes first!"* This concept relates to presenting food that is visually appealing, and it also impacts portion size and pricing.

A pasta entrée filling an 8-inch plate may be lost on an 11-inch plate. However, guests receiving the pasta entrée on an 8-inch plate will likely perceive higher levels of value than those receiving the same entrée on an 11-inch plate, even though portion size and cost to the operator is almost identical.

Portion size is a function of both product quantity *and* presentation. Many successful cafeterias use smaller than average dishes to plate their food. For their guests, the image of price to value when dishes appear to be full comes across loud and clear.

In some foodservice operations and particularly in "all you care to eat" facilities, the previously mentioned principle again holds true. The proper dish size is just as critical as the proper sized scoop or ladle when serving the food. Of course, in a traditional table service operation, an operator must carefully control portion sizes because the larger the portion size, the higher the product costs.

Many of today's health-conscious consumers prefer lighter food with more fruit and vegetable choices. The portion sizes of these items can often be boosted at a low increase in product cost. At the same time, average beverage sizes are increasing as are portion sizes of many side items including French fries. If these items are lower-cost items, this can be good news for the operator. However, it is still important to consider the costs of larger portion sizes.

Every menu item to be priced should be analyzed to help determine if the quantity (portion size) being served is the "optimum" quantity. Operators would, of course, like to serve that amount, but no more. For operators managing their cost control efforts, the effect of portion size on menu prices is significant, and back-of-house staff should establish and maintain careful control over an operation's desired portion sizes.

Find Out More

Foodservice operators have a variety of choices when selecting software programs designed to help calculate the cost, and thus the appropriate pricing, of their various menu items. These menu management programs allow operators to insert their standardized recipes, portion sizes, and the cost of the ingredients used to make the recipes, and then the items' portion costs are automatically calculated.

Advances in menu management software continue to occur rapidly. Increasingly, foodservice operators look for programs providing pricing options and then design their own pricing and sales tracking processes.

To stay current with newly developed menu pricing software and related apps, enter "menu item pricing software" in your favorite search engine and review the results.

Delivery methods have become an increasingly important factor in establishing menu prices. There are essentially four ways foodservice operators can deliver purchased menu items to their guests:

1) Dine-in service: Menu items are delivered to guests at their on-site tables or other seating areas.
2) Pick up/carryout: Guests receive their menu items from drive-through windows or pick up ordered items from designated on-site carry out areas.
3) Operator direct delivery: Menu items are delivered to guests by an operation's own delivery employees.
4) Third-party delivery: Menu items are delivered to guests by third-party delivery (see Chapter 5) partners selected by the operation).

The delivery style with the greatest impact on menu prices is third-party delivery. From the perspective of guests, the decision to utilize a third-party delivery company such as Grubhub, DoorDash, or Uber Eats means the guest is placing important value on convenience as well as on the menu items they choose.

When guests order directly from a third-party delivery company, they pay for:

✓ The selling price of their selected menu items
✓ A service fee charged by the delivery company for providing the service
✓ A delivery fee for the food that is delivered
✓ A gratuity; an optional tip for the delivery driver

The COVID-19 pandemic era saw explosive growth in third-party delivery companies as many restaurants either closed indoor dining areas or severely restricted their inside seating capacities. Third-party delivery services are popular with many guests; however, the services of third-party delivery companies are not free to restaurants.

Depending upon the specific arrangement made between a foodservice operation and its third-party delivery partner(s), the operation will pay between 10 and 30% of a guest's total bill to the third-party delivery company. For example, if a foodservice operation has a customer who utilizes a third-party delivery app and purchases $100 worth of menu items from the operation, the operation will receive only $70.00 to $90.00 from the third-party delivery company.

The foodservice operation typically pays a significant fee to satisfy their customers' desire for convenience. Some foodservice operators charge different (higher) menu prices when items are ordered from a third-party delivery app, or they avoid the use of third-party delivery services entirely.

Meal Period

In some cases, guests willingly pay more for an item served in the evening than for that same item served during a lunch period. Sometimes this is the result of a smaller "luncheon" portion size or some different menu items. However, in other cases the portion size and service levels may be the same in the evening as earlier in the day. This is true, for example, in buffet restaurants that charge a different price for lunch than for dinner. Perhaps operators expect those on lunch break to spend less time in the operation, and they will then eat less. Alternatively, they may believe their guests simply seek to spend less for lunch than dinner.

Foodservice operators must exercise caution in this area. Guests should clearly understand why a menu item's price changes with the time of day. If this cannot be answered to the guest's satisfaction, it may not be wise to implement a time-sensitive pricing structure.

Location

Some foodservice operators make their locations a major part of their key pricing decisions. This is illustrated, for example, by food facilities operated in themed amusement parks, movie theaters, on golf courses, and within sports arenas.

Foodservice operators in these types of locations can charge premium prices because they have a monopoly on food sold to visitors. The only all-night restaurant on an interstate highway exit is in much the same situation. Contrast that with an operator who is just one of 10 similar seafood restaurants on a **restaurant row** in a seaside resort town. In this case it is unlikely that one of the ten operations can charge prices significantly higher than its competitors based solely on location.

Key Term

Restaurant row: A street or region well-known for having multiple foodservice operations a close distance from each other.

Operators should not discount the value of an excellent restaurant location, and location alone can influence price in some cases. Location does not, however, guarantee long-term success. Location can be an asset or a liability. If it is an asset, menu prices may reflect that fact. If location is a liability, menu prices may need to be lower to attract a sufficient clientele to ensure the operation achieves its total revenue and profit goals.

Bundling

Bundling refers to the popular practice of selecting specific menu items and pricing them as a group (bundle) so the single menu price of all the items purchased together is lower than if the items in the group were purchased individually.

A common example of bundling is the combination meals offered by quick-service restaurants. In many cases, these bundled meals consist

Key Term

Bundling: A pricing strategy that combines multiple menu items into a grouping which is then sold at a price lower than that of the bundled items purchased separately.

of a sandwich, French fries, and a drink. Bundled meals, are often promoted as "combo meals" or "value meals," and are typically identified by a number (e.g. Number 6 or Number 7) or a single name (e.g. Everyday Value Meal (Arby's), Mix and Match (Burger King), and Cheeseburger Combo (McDonalds) for ease of ordering.

Bundled menu offerings are carefully designed to encourage guests to buy all menu items included in the bundle, rather than to separately purchase only one or two of the items. Bundled meals are typically priced very competitively, and a strong value perception can be established in the guest's mind.

Find Out More

You have now been introduced to several factors that can influence the prices foodservice operators charge for the items they sell. In the future, Leadership in Energy and Environmental Design (LEED) certification achieved by an operation may well constitute another such factor.

The LEED rating system developed by the U.S. Green Building Council (USGBC) evaluates facilities on several standards. The rating system considers sustainability, water use efficiency, energy usage, air quality, construction and materials, and innovation.

Increasingly, many consumers are willing to pay *more* to dine in LEED-certified operations. In addition, LEED-certified buildings are healthier for workers and for diners. The LEED certification creates benefits for foodservice operators, employees, and guests, and it will likely continue to be of increasing importance to guests.

To learn more about LEED-certification in the foodservice industry, enter "LEED-certified restaurant standards" in your favorite search engine and view the results.

What Would You Do? 7.1

"We have to lower our prices because there is nothing else we can do!" said Ramon, Director of Operations for the seven-unit Boston Sub Shops. Boston's was known for its modestly priced but very high-quality sandwiches and soups.

Business and profits were good, but now Ramon and Judy, who was one of the company's store managers, were discussing the new $8.99 "*Foot Long Deal*" sandwich promotion just announced by their major competitor: an extremely large chain of sub shops that operates thousands of units nationally and internationally.

"They just decided to lower their prices to appeal to value-conscious customers," said Ramon.

"But how can they do that and still make money?" asked Judy.

"There's always a less expensive variety of ham and cheese on the market," replied Ramon. "They use lower-quality ingredients than we do, and we charge $10.99 for our foot-long sub. That wasn't bad when they sold theirs at $9.99. Our customers know we are worth the extra dollar. Now they are running their special at $8.99. I don't know, but this might really hurt us," said Ramon, "What should we do?"

Assume you were Judy. How do you think your guests will respond to this competitor's new pricing strategy? What specific steps would you recommend to Ramon and Judy that can help Boston's Subs address this new pricing/cost challenge?

Methods of Food and Beverage Pricing

Many factors impact how a commercial foodservice operation establishes its prices, and the methods used are often as varied as the operators who use the methods.

Menu item prices are directly affected by one or more of the factors previously described. However, menu prices have historically been determined based on either an operation's menu item costs or desired profit level per menu item. This makes sense when the operator's perspectives about price formula introduced earlier in this chapter are re-examined closely:

Item Cost + Desired Profit = Selling Price

When foodservice operators focus on a menu item's cost to establish price, they generally recognize that items costing more to produce must be sold at prices higher than lower cost items. The actual prices charged in a foodservice operation are primarily determined by its owner, perhaps with assistance from the operation's managers and production staff. However, those primarily responsible for successfully pricing an operation's menu items must have a good understanding of both the cost-based and profit-based approaches to pricing.

Cost-based Pricing

When using cost-based pricing a foodservice operator calculates the cost of the ingredients required to produce the menu item being sold. The cost of the menu item includes all menu item accompaniments. For example, when a foodservice operation sells a dinner entree including a salad and dinner rolls at one price, the entrée's cost must also include that of the salad, dressing, and dinner rolls and butter included with the entree.

In most cases, cost-based pricing is based on the idea that the cost of producing an item should be a predetermined percentage of the item's selling price. With this pricing system, menu items having a lower food cost percentage when compared to their selling prices are typically considered more desirable than menu items with higher percentage costs.

Key Term

Food cost percentage: The portion of food revenues spent on food expenses.

The formula used to compute a menu item's food cost percentage is:

$$\frac{\text{Item food cost}}{\text{Selling price}} \times 100 = \text{Food cost percentage}$$

To illustrate, if it costs an operator $4.00 to purchase the ingredients needed for a menu item, and the item is sold for $20.00, the item's food cost percentage is calculated as:

$$\frac{\$4.00 \text{ item food cost}}{\$20.00 \text{ selling price}} = .20 \text{ or } 20\%$$

When operators utilize food cost percentage as a major factor in pricing their menu items, the cost of producing one portion of the menu item must be accurately determined. When the item sold is a single portion item (e.g. one New York strip steak), the cost of producing one portion is relatively straightforward.

However, when menu items are produced in multiple portions (example: a pan of lasagna containing 12 portions is prepared), operators must calculate their **portion cost** based on the standardized recipe (see Chapter 1) used to produce the menu item.

Key Term

Portion cost: The product cost required to produce one serving of a menu item.

The use of standardized recipes is critical for operators utilizing product cost as a primary determinant of menu prices. The food cost of producing a menu item must be known, and this cost must be consistently the same if selling prices are based on product costs. Also, when ingredient costs change, the increased cost must be used to calculate the new portion costs resulting from standardized recipes that now contain higher cost ingredients. These new portion costs may (or may not) cause an operator to change menu prices, but the actual costs should be known.

While for-profit foodservice operations are concerned with food cost percentages, non-profit operations are typically more interested in the cost to serve each guest.

Examples of non-profit operations in which cost per meal is important to operators include military bases, correctional facilities, hospitals, senior living facilities, schools and colleges, and large business organizations that offer foodservices to their employees.

Whether the individuals served are soldiers, inmates, patients, residents, students, or employees, calculating an operation's cost per meal is easy because it uses a variation of the basic food cost percentage formula:

$$\frac{\text{Cost of Food Sold}}{\text{Total Meals Served}} = \text{Cost Per Meal}$$

For example, assume a non-profit operation incurred $65,000 in cost of food sold during an accounting period. In the same accounting period, the operation served 10,000 meals. To calculate this operation's cost per meal, the cost per meal formula is applied.

In this example it would be:

$$\frac{\$65,000 \text{ cost of food sold}}{10,000 \text{ meals served}} = \$6.50 \text{ per meal}$$

Whether managers are most interested in their food cost percentage or in their cost per meal served, it is essential that they first accurately calculate their product costs. When foodservice operators have established a target food (or beverage) cost percentage they can use it to determine menu item prices.

For example, if a menu item has a food cost of $8.00 and an operation's desired cost percentage is 40%, the following formula can determine the item's menu price:

$$\frac{\text{Food Cost of Menu Item}}{\text{Desired Food Cost \%}} = \text{Selling Price}$$

In this example:

$$\frac{\$8.00 \text{ Food cost of menu item}}{40\% (.40) \text{ Desired food cost}} = \$20.00 \text{ Selling Price}$$

Another method of calculating selling prices based on predetermined product cost percentage goals uses a pricing factor (multiplier) assigned to each potentially desired food or beverage cost percentage. This factor, when multiplied by an item's portion cost indicates a selling price that yields the desired product cost percentage. Some commonly used pricing factors are presented in Figure 7.2.

The above pricing factor method of establishing menu prices is easy to use. For example, if an operator wants a 25% product cost, and a menu item has a food cost of $4.50, the following pricing formula would be used:

Item Food Cost × Pricing Factor = Selling Price

Desired Product Cost %	Pricing Factor
20	5.000
23	4.348
25	4.000
28	3.571
30	3.333
$33^1/_3$	3.000
35	2.857
38	2.632
40	2.500
43	2.326
45	2.222

Figure 7.2 Pricing Factor Table

In this example that would be:

$4.50 item food cost × 4.0 pricing factor = $18.00 selling price

The two methods to determine a proposed selling price based on product cost percentage yield identical results. With either approach, an item's selling price is determined with the goal of achieving a specified food or beverage cost percentage for each item sold.

Contribution Margin-based Pricing

Some commercial foodservice operators use a profit-based rather than cost-based pricing method to establish menu item selling prices. These operators set their prices based on the menu item's **contribution margin (CM)**.

CM for a single menu item is defined as the amount of money that remains after the product cost of the menu item is subtracted from the item's selling price. It is, then, the dollar amount that a menu item "contributes" to pay for labor and all other expenses and to the operation's profits.

Key Term

Contribution margin (CM): The dollar amount remaining after subtracting a menu item's product costs from its selling price.

To illustrate CM pricing, assume a menu item sells for $18.75 and the cost of the food to produce the item is $7.00. In this example the CM for the menu item is calculated as:

Selling Price − Item Food Cost = Contribution Margin (CM)

or

$18.75 selling price − $7.00 item food cost = $11.75 CM

When this approach is used, the formula for determining a menu item's selling price is:

Item Food Cost + Contribution Margin (CM) Desired = Selling Price

When using the CM approach to establish selling prices, operators most often develop different CM targets for various menu items or groups of items. For example, in an operation where items are priced separately, entrées might be priced with a CM of $10.50 each, desserts with a CM of $5.25 each, and non-alcoholic drinks with a CM of $3.75.

To apply the CM method of pricing, foodservice operators use a two-step process.

Step 1: Determine the average contribution margin required for each item
Step 2: Add the contribution margin required to the item's product cost

Step 1: Operators determine the average CM they require based on the number of items to be sold or on the number of guests to be served. The process used for each approach is identical.

For example, to calculate CM based on the number of items to be sold, operators add their non-food operating costs to the amount of profit they desire, and then divide the result by the number of items expected to be sold:

$$\frac{\text{Nonfood costs} + \text{Profit desired}}{\text{Number of items to be sold}} = \text{CM desired per item}$$

To calculate CM based on the number of guests to be served, operators divide all nonfood operating costs, plus the amount of profit they desire, by the number of expected guests:

$$\frac{\text{Non-food costs} + \text{profit desired}}{\text{Number of guests to be served}} = \text{CM desired per guest served}$$

For example, if an operator's budgeted non-food operating costs for an accounting period are $125,000, desired profit is $15,000, and the number of items estimated to be sold is 25,000, the operator's desired average CM per item are calculated as:

$$\frac{\$125,000\,(\text{nonfood costs}) + \$15,000\,(\text{desired profit})}{25,000\,(\text{Number of items to be sold})} = \$5.60\,\text{CM desired per item}$$

Step 2: Operators complete this step by adding their desired CM per item (or guest) to the cost of preparing a menu item. For example, if an operator's desired average CM per item is $5.60 and a specific menu item's food cost is $3.40, the item's selling price would be calculated as:

$3.40 item food cost + $5.60 CM desired + = $9.00 selling price

The CM method of pricing is popular because it is easy to use, and it helps ensure that each menu item sold contributes to an operation's profits. When using CM to set menu prices, the prices charged for menu items vary only due to variations in product cost. When managers have accurate budget information about their nonfood costs taken from recent income statements and realistic profit expectations, the use of the CM method of pricing can be very effective.

Operators who utilize the CM approach to pricing do so believing the average CM per item sold is a more important consideration in pricing decisions than the product cost percentage. The debate over the "best" pricing method for food and beverage products is likely to continue. However, all operators must view pricing as an important process, and its goal is to consider a desirable price/value relationship for guests.

In the final analysis, the customer eventually determines what an operation's total sales will be for each menu item. Experienced operators know that sensitivity to required profit and to the guests' needs, wants, and desires are the most critical components of an effective pricing strategy.

Technology at Work

Most food service operators establish menu prices using their preferred process and then continue to use these prices until they next review their costs or reprint their menus. A shift to digital ordering, combined with higher food and labor costs, however, has led some foodservice operators to adopt dynamic pricing strategies.

Essentially, dynamic pricing is the process of adjusting selling prices based on consumer demand, rather than costs alone. In most cases, higher guest demand means that higher prices are charged. The concept of dynamic pricing is not new. First initiated by airline companies, today dynamic pricing is common in hotel chains, amusement parks, golf courses, sports venues, and other service industry businesses. Since increasing numbers of foodservice operators' menus are created and displayed in digital form, temporarily changing prices based on consumer demand is much easier than in previous times (when a change in price required a complete menu reprint).

In most cases, the foodservice industry has been slow to aggressively implement dynamic pricing. For example, consider a foodservice operator selling a Ribeye steak on a busy Saturday night when the dining room is full, and there is a long wait list. Many foodservice operators' will sell that same steak, at the same price, on a Tuesday evening when business is much slower. It is interesting to note, however, that food service operators have long used dynamic pricing in reverse. That is, it is common for a foodservice operation to offer *reduced* drink prices during "Happy Hour," or to offer "Early Bird" menu specials sold at reduced prices during early meal times (when business is slow).

It remains to be seen how aggressive foodservice operators may become with dynamic pricing. There is no doubt, however, that it is an area that should be monitored carefully by professionals in the foodservice industry. To learn more about how some foodservice operators are selectively utilizing dynamic pricing, enter "dynamic pricing in restaurants" in your favorite search engine and review the results.

Evaluation of Pricing Efforts

Regardless of the method used to establish selling prices, foodservice operators should regularly evaluate the results of their pricing efforts, and many operators evaluate menu items based on two key characteristics: popularity and profitability.

Menu engineering is a term popularly used to describe one method that addresses and examines these two variables.

Operators using the menu engineering process desire to produce a menu maximizing its overall CM. To use menu engineering operators must sort their menu items by two variables:

Key Term

Menu engineering: A system used to evaluate menu pricing and design by categorizing each menu item into one of four categories based on profitability and popularity.

1) Popularity (number of each item sold)
2) Weighted contribution margin

Calculating Menu Item Popularity (Number Sold)

To calculate the average popularity (number sold) of a menu item, operators must first determine the average number of all items sold:

$$\frac{\text{Total Number of Menu Items Sold}}{\text{Number of Menu Items Available}} = \text{Average Number Sold}$$

For example, if an operator sold 5000 entrees during a specific period, and the operator's menu lists 10 different entree choices, the average popularity of the entrees sold during that time period would be calculated as:

$$\frac{5000\ \text{Menu Items Sold}}{10\ \text{Menu Items Available}} = 500\ \text{Menu Item Average Popularity (Number Sold)}$$

When using menu engineering and applying the 500 average popularity, any menu item that sold *more* than 500 times during the analysis period is classified as "High" in popularity, and menu items selling *less* than 500 times would be classified as "Low" in popularity.

Calculating Weighted Contribution Margin

To continue the menu engineering process operators must also define the weighted average CM of their menu items. Some operators confuse averages (means) with weighted averages. However, the distinction between the two is important.

To use a simple example, assume an operator collected the following data and wanted to calculate the average size of the sale made in their operation over a 3-day reporting period.

Week Day	Guests Served	Total Sales	Average Sale
Monday	50	$ 500	$10.00
Tuesday	150	$1,650	$11.00
Wednesday	250	$3,000	$12.00
Total/Average	450	$5,150	?

In this example, to calculate the size of the "average" sale on Monday through Wednesday, one should NOT use the unweighted formula that is typically used to calculate a mean (average). That *unweighted formula* is:

$$\frac{\$10.00 + \$11.00 + \$12.00}{3\,days} = \$11.00 \text{ per day average}$$

In fact, what the operator really wants to learn when calculating the average sales for the three days is "*How much did the average guest spend in my operation from Monday through Wednesday?*"

The number of guests served each day varied, so the operator determines a *weighted* average sale formula as shown below:

$$\frac{\$5,150 \text{ (total sales during all 3 days)}}{450 \text{ (total guests served in all 3 days)}} =$$
$11.44 weighted average sale per guest

Note that the operator's average sale size resulting from using unweighted and weighted average formulas in this example differ.

Returning to menu engineering, to calculate the average **weighted contribution margin** for their menu items, operators must first calculate

Key Term

Weighted contribution margin: The contribution margin provided by all menu items divided by the total number of items sold. Weighted contribution margin is calculated as:

$$\frac{\text{Total Contribution Margin of All Items Sold}}{\text{Total Number of Items Sold}} =$$
Weighted contribution margin

the total contribution margin generated by all items sold, and then divide by the number of items sold (see Figure 7.3).

Column A in Figure 7.3 lists the name of individual menu items. Column B lists the total number of each menu item sold. Note that the sales (popularity) of items vary from a low of 190 sold (Item 10) to a high of 1050 sold (Item 4).

Column C lists the individual CM of each of the 10 items offered for sale, and column D lists the total item CM generated by each menu item.

The value in Column D is calculated by multiplying the value in Column B times the value in Column C. In this example the average number sold is 500, and the average weighted CM is $12.43 ($62,174 total item CM / 5000 sold = $12.43.)

After an operator has calculated the popularity and weighted contribution margin of the items listed on the menu, the items are sorted into a 2 × 2 menu engineering matrix containing four squares as shown in Figure 7.4.

Figure 7.4 shows that menu items with sales above the average level of popularity (500 sold in this example) are "High" in popularity, and items that sold less than 500 times are "Low" in popularity. Similarly, those menu items whose contribution margins are above the weighted contribution margin average ($12.43 in this example) are "High" in contribution margin, and items with a lower contribution margin are "Low" in contribution margin.

Column A	Column B	Column C	Column D
Menu Item	Total Number Sold	Single Item Contribution Margin	Total Item Contribution Margin
1	250	$ 14.50	$ 3,625
2	250	$ 7.50	$ 1,875
3	525	$ 12.50	$ 6,563
4	1050	$ 17.25	$18,113
5	510	$ 7.00	$ 3,570
6	625	$ 13.50	$ 8,438
7	400	$ 12.75	$ 5,100
8	825	$ 10.50	$ 8,663
9	375	$ 8.25	$ 3,094
10	190	$ 16.50	$ 3,135
Total	5000	$120.25	$62,174
Average (Mean)	500	$ 12.03	
Weighted contribution margin			$ 12.43

Figure 7.3 Total Contribution Margin Worksheet for 10 Item Menu

Popularity

		Low	High
	High	High-contribution margin Low popularity PUZZLE	High-contribution margin High popularity STAR
Contribution Margin	Low	Low-contribution margin Low popularity DOG	Low-contribution margin High popularity PLOW HORSE

Figure 7.4 Menu Engineering Matrix

Many users of menu engineering name the items contained in the four squares for ease of remembering the characteristics of each item. These commonly used names (Puzzles, Stars, Dogs, and Plow Horses) are also shown in Figure 7.4.

Figure 7.5 shows where each of the ten example menu items listed in Figure 7.3 would be located.

Menu Modifications

Operators should regularly analyze their menus to make needed modifications and improvements. When using menu engineering, each menu item that fell within the four squares requires a special reaction from an operator. Examples of suggested menu modification strategies resulting from menu engineering are summarized in Figure 7.6.

One reason to perform menu engineering analysis is to identify items whose prices must or can be increased to enhance an operation's profitability. Some

Popularity

		Low	High
Contribution Margin	High	PUZZLE Menu items 1, 7, and 10	STAR Menu items 3, 4, and 6
	Low	DOG Menu items 2 and 9	PLOW HORSE Menu items 5 and 8

Figure 7.5 Menu Engineering Results

operators are hesitant to raise prices fearing customers will react negatively. Experienced foodservice operators, however, know that price is not the only determining factor when guests decide where to spend their dining out dollars.

Item	Characteristics	Problem	Management Action
Puzzle	High contribution margin; Low popularity	Marginal due to lack of sales	a) Relocate on menu for greater visibility. b) Consider reducing the selling price.
Star	High contribution margin; High popularity	None	a) Promote well. b) Increase prominence on the menu.
Dog	Low contribution margin: Low popularity	Marginal due to low contribution margin and lack of sales	a) Remove from menu. b) Consider offering as a special occasionally, but at a higher menu price.
Plow Horse	Low contribution margin; High popularity	Marginal due to low contribution margin	a) Increase menu price. b) Reduce prominence on the menu. c) Consider reducing portion size.

Figure 7.6 Potential Menu Modifications

Technology at Work

A foodservice operation's menu is much more than a list of foods and beverages. It can be a powerful marketing tool to familiarize customers with an operation. It can also get them excited about the unique items an operation offers for sale.

Regularly evaluating individual menu items for popularity and profitability is an important task because it helps identify both profit producing items and those that perform (sell) poorly. With this information known, menus can be modified to optimize sales and profits.

Fortunately, there are several useful software programs available that help operators perform menu engineering analysis. To review the features and costs of such programs enter "Menu engineering software" in your favorite search engine and view the results.

The quality of customer service, cleanliness, staff friendliness, and uniqueness of menu items can be *more* important than price. The best operators ensure that these aspects of their operations meet or exceed guests' expectations. Then price increases acceptable to guests that help ensure an operation's long-term profitability can be implemented.

Foodservice operators who design menus and price their menu items properly are in an excellent position to successfully achieve operating goals. To achieve those goals, however, operators will need a dedicated team of professionals to help them serve guests in a way that makes them want to come back. Recruiting, selecting, and training team members who have, or can be taught, the skills needed to reach all the operation's financial and service-related goals is important, and will be the sole topic of the next chapter.

What Would You Do? 7.2

"$39.95—that's over ten dollars more than we charged for it yesterday!" said Shilo, the Dining Room manager at Chez Franco's restaurant.

Shilo was discussing the day's dinner menu with Andrea, the restaurant's executive chef. Andrea had just shown Shilo the daily menu insert his service staff would use that night. On the night's new menu he noticed that the price of Pontchartrain Style Snapper, one of the operation's most popular dishes, had increased overnight. Yesterday it sold for $27.95. Today Andrea had priced it at $39.95.

"Tell me about it," replied Andrea, "our seafood supplier really raised the price on our latest delivery. My red snapper cost is now up by almost $11.00 per pound. That's over $4.00 a portion. The supplier said there was a snapper shortage, and he wasn't sure how long it would last. With the new cost of snapper, I needed this price increase to keep our food cost percentage in line.

Shilo wasn't sure his servers or the guests to be served that night would be very happy with Andrea's pricing decision. The snapper was a very popular item, and that meant tonight lots of customers would likely notice the price increase and make a comment about it!

Assume you were Shilo on the night this new menu price was initiated. How would you likely respond to a returning guest who questioned the significant price increase on the red snapper item? What do you think you should tell your servers to say in response to guests' anticipated reactions to this menu item's price increase? What else might you do to address this issue?

Key Terms

Price (noun)	Wagyu Beef	Contribution
Price (verb)	Restaurant row	margin (CM)
Value proposition	Bundling	Menu engineering
Profit margin	Food cost	Weighted contribution
Consumer rationality	percentage	margin
Law of demand	Portion cost	

Operator's 10-Point Tactics for Success Checklist

Evaluate your need for, and the current status of, each of the following operational tactics. For those tactics you think are important, but not yet in place, develop an action plan for its implementation including who will be responsible for the tactic's completion and the target date by which it should be completed.

Tactic	Don't Agree (Not Done)	Agree (Done)	Agree (Not Done)	If Not Done — Who Is Responsible?	Target Completion Date
1) Operator understands the importance of price in the profitable operation of a foodservice business.	____	____	____		
2) Operator has carefully considered the difference between an operator's view of price and guests' views of prices.	____	____	____		
3) Operator has considered the impact of economic conditions, local competition, level of service, and guest type when establishing menu prices.	____	____	____		
4) Operator has considered the impact of product quality, unique features, portion size, delivery style, meal period, and location when establishing menu prices.	____	____	____		

(Continued)

Tactic	Don't Agree (Not Done)	Agree (Done)	Agree (Not Done)	Who Is Responsible?	Target Completion Date
				If Not Done	
5) Operator understands the concept of bundling when establishing menu prices.	____	____	____		
6) Operator has considered the value of utilizing a product cost-based approach when establishing menu prices.	____	____	____		
7) Operator has considered the value of utilizing a contribution margin-based approach when establishing menu prices.	____	____	____		
8) Operator understands the importance of analyzing the menu as a means of better understanding guests' purchasing preferences.	____	____	____		
9) Operator understands how to use menu engineering to analyze menus.	____	____	____		
10) Operator recognizes the importance of carefully implementing any menu price adjustments to minimize guest dissatisfaction and to optimize sales.	____	____	____		

8

Successful Employee Staffing and Training

What You Will Learn

1) How to Recruit Team Members
2) How to Select Team Members
3) Required Documentation for New Team Members
4) How to Train Team Members

Operator's Brief

In this chapter, you will learn that, in today's tight and competitive labor market, obtaining and retaining an efficient workforce will be one of your most significant challenges. Attracting and retaining an effective work team begins with successful recruiting, and ways to do this include internal recruiting, traditional external recruiting, and online recruiting. This chapter addresses each of these.

Some of your most important staffing decisions will relate to the actual selection of new team members. In this chapter you will learn about several selection-related tools. These include formal job applications, employment interviews and, sometimes, pre-employment testing and background checks. Using these tools properly and legally is important, and the chapter addresses how to do so.

After new employees are selected, employment documentation is required. The chapter addresses documentation at the federal, state, and, when applicable, local levels.

In this chapter, you will learn that building effective work teams requires careful planning and training programs. Experienced foodservice operators know successful trainers must possess specific personal characteristics, and in this chapter, you will learn about these factors and why they are important.

(Continued)

The actual creation of an effective training program is a five-step process, and these steps must occur in the sequence cited below:

Step 1: Identify Training Needs
Step 2: Develop Training Objectives
Step 3: Develop Training Plans
Step 4: Create Training Lessons
Step 5: Develop (and update) Training Handbooks

Regardless of the method used to deliver training, evaluation of training efforts is essential. Training evaluation that occurs before, during, and after training is critical to ensure that expenditure of training resources represents the best use of these funds.

CHAPTER OUTLINE

Recruiting Team Members
    Internal Recruiting
    Traditional External Recruitment Methods
    Internet-based External Recruitment Methods
Choosing Team Members
    Applications
    Interviews
    Pre-employment Testing
    Background Checks and References
Required Documentation for New Team Members
    Form I-9 (Federal Form)
    Form W-4 (Federal Form)
    State and Local Tax Forms
Training Team Members
    Benefits of Training
    Characteristics of Successful Trainers
    Steps in Creating Effective Training Programs
    Evaluation of Training Efforts

Recruiting Team Members

Foodservice operators in all industry segments and all sizes face a variety of challenges as they attempt to successfully achieve their operational goals. To reach those goals, foodservice operators must serve excellent products and provide excellent guest service. But, to do so, they must recruit, train, and retain a team of dedicated professionals to help them. Today, the number one challenge to success that foodservice operators most consistently report is that of obtaining and maintaining a sufficient number of qualified staff members.

Operators facing staffing challenges can best help themselves and businesses when they consider the four factors that will make them most competitive in the labor market. These are:

1) Compensation
2) Benefits
3) Scheduling
4) Working conditions

1) Compensation

The compensation required to be competitive for recruiting and retaining workers varies based on a foodservice operation's physical location. Those businesses located in high cost of living areas typically must offer greater pay for the same work than do businesses located in lower cost of living areas. Similarly, in areas where the supply of labor is limited, operators will find that higher wages are required to attract workers.

It is also true, however, that an employer need not offer the *highest* wages in an area to be considered the area's **employer of choice**. Competitive pay is important, but it is often *not the most important factor* in securing talented foodservice workers. To address the issue of appropriate pay, foodservice operators should survey their competitors, consider the economic conditions in their area, and then establish pay rates or pay ranges that are competitive for each position.

Key Term

Employer of choice: A company whom workers choose to work for when given employment choices. This choice is a conscious decision made when initially joining a company and when deciding to stay with an employer.

2) Benefits

U.S. Labor laws require that all employers offer their workers some **mandatory benefits**. Under current federal and state laws mandatory these benefits include:
- Social Security and Medicare
- Unemployment insurance
- Workers' compensation insurance
- Family and Medical Leave Act (FMLA) protections

Key Term

Mandatory benefit: An employee benefit that must, by law, be paid by an employer.

Key Term

Voluntary benefit: An employee benefit paid at the discretion of an employer in efforts to attract and keep the best possible employees.

In addition to mandatory benefits, nearly every foodservice operation offers some level of **voluntary benefits** to its employees. On the very lowest end of the scale, this may include offering free or reduced price meals during working shifts. On the other end of the scale, employers may offer health insurance, paid vacation time, and the matching of 401(k) contributions.

Regardless of the level of benefits offered, it is important for operators to recognize that benefits will be a significant determining factor in many employees' decisions about whether to accept or decline an employment offer. This is especially true when pay rates or pay ranges offered for a specific job in an area are relatively similar.

To illustrate, consider the foodservice operator seeking to hire a tipped server at a rate of $10 per hour which is the competitive wage rate for servers in a specific area. This foodservice operator offers only reduced-cost meals to employees. However, a competitive operator also offers a $10 per hour rate of pay. However, that potential employer includes a free employee meal and a $200 per-month contribution toward a health-care plan and two weeks' paid vacation after one year of service. It is easy to see that the applicant in this scenario will likely accept the greater benefits package. In fact, a high-quality employee may decide to work for the competitor operator even if the hourly rate of pay offered to them was only $9.50 (or less) per hour!

3) **Scheduling**

The work schedules (see Chapter 10) assigned to foodservice employees are particularly important in their employment decisions, and this is especially so for those who work part-time. Potential employees can choose to work part-time for several reasons including:

- Childcare issues: These workers choose to work part-time for reasons related specifically to affordable, available, and/or appropriate childcare.
- Personal obligations: These can include family or home-related reasons including staying at home with a sick child or parent, or to do housework or another chore for someone.
- Health or medical issues: The worker's own illness, injury, or disability may prevent the employee from working full time.
- School or training: A staff member may work part-time to attend high school, college, or another training program.
- Retired and /or Social Security earnings limit: Some employees work part-time because they are retired or cannot work more hours without losing a portion of Social Security benefits.

It is easy to see that in all these cases (and more), the specific work schedule that an employer offers a potential employee can make a great deal of difference in the willingness of the potential employee to accept a job. It is also true that a work schedule acceptable to one part-time employee may not be acceptable to another staff member. Therefore, it is generally important to allow maximum scheduling flexibility in the jobs that are offered.

Some foodservice operators feel they do not have any flexibility regarding employee scheduling. They may maintain that employees are needed when required work must be performed, so scheduling flexibility is not an option. However, in many cases these operators are wrong.

To illustrate, consider the operator whose bar station is particularly busy on Friday and Saturday nights. While the operator normally can fully staff the bar with one bartender, on Friday and Saturday nights the operator needs two bartenders. The operator has proposed a shift of 6:00 p.m. to 2:00 a.m., to cover the time the bar is busy.

If the operator creates a part-time bartender's job (two weekend nights per week with 8 hours on Saturday and Sunday) the operator will have extreme difficulty filling this position.

Reasons may include that few part-time workers want to work every Friday and Saturday night, and potential employees who work full-time Monday through Friday jobs may also not want to work over the weekend.

In this example it might be more sensible for the operation to provide bartender training to one or more employees in server positions. Then these employees can be scheduled on the Friday and/or Saturday nights that have busy bartending shifts. In this way, proper scheduling will allow the operator to fill the bartender's position with qualified staff, an otherwise difficult task.

4) Working Conditions

For many lower-waged foodservice workers, pay rates, benefits, and schedules offered by potential employers may be similar. When they are, working conditions may become the most important factor when selecting one employer over another and in deciding to stay with that employer over an extended time.

If they were asked, most foodservice operators would likely say they want to provide their workers with a *"good"* place to work. Successful foodservice operators recognize their responsibility to provide a healthy, safe, and secure work environment. They also create policies and procedures that ensure their employees are as free as possible from risks posed by the workplace, other employees, and guests. However, this recognition is still not typically sufficient to ensure an operation can attract and retain all necessary workers.

Tips for creating a welcoming and supportive work environment are numerous, and effective foodservice operators can choose from those that apply most directly to them. There are, however, some no-cost characteristics of a positive workplace that all foodservice operators can provide to ensure working conditions are welcome. Among the most important of these are:

a) Demonstration of genuine respect

Respect for all team members is the foundation of a positive work environment. Every foodservice worker should be treated with respect, and every employee must always treat others in the same way. Regardless of the tasks they are assigned, every foodservice worker wants to know they are valued, appreciated, and will be treated fairly. In many cases when foodservice workers are asked why they left a previous employer, the answer is not low pay or a dislike of the actual work. Rather, it is frustration resulting from not being treated in a manner the employee felt was most appropriate.

Improper comments, gestures, and actions can be initiated by management, co-workers, and even guests. It is a major responsibility of foodservice operators who want to ensure their employees join and stay with organization that regularly confirm the worth of each worker. This can be done as easily as verbally welcoming employees as they arrive at work and thanking them for a job well done as they depart from their jobs.

b) Open communication

Workers in every foodservice operation experience times of stress and frustration. Sometimes this results from the work itself. For example, when a foodservice operation is very busy the demands on servers and production staff can be high. While stress of this type may not be avoidable, it is possible for all employees to feel like they have a voice at work. Therefore, it is important that operators encourage feedback from these employees. **Open-door communication** policies are nearly always present in operations that recognize the importance of the policies.

From its definition, one might think an open-door communication policy is passive, and operators must wait for employees to initiate communication. Not true!

> **Key Term**
>
> **Open-door communication (policy):** A policy that indicates supervisors or managers are willing to discuss an employee's questions, complaints, suggestions, and concerns at convenient times.

An open-door policy implies that communication goes both ways. It can be initiated by foodservice operators who inquire about workers' issues and by staff members who want to discuss job-related issues with the operators.

c) A sense of teamwork

Many management topics have been written about the importance of teamwork to an organization's success. Often authored by sports personalities and/or successful business persons, there is most often little disagreement with the statement that *"Teamwork makes the dream work."*

It is true that teamwork is essential to success. However, it is also essential to create positive work environments. These are critical in employees' efforts to select an employer and stay with that organization because they want to be part of a cohesive and welcoming work team.

One often neglected aspect of team building involves addressing team members who are not productive and may even be disruptive to the team (see Chapter 10). Some supervisors may attempt to change these employee's attitudes, and doing so may sometimes be a worthwhile effort. However, foodservice operators must also recognize those times when specific team members must be removed from the team. This can be particularly challenging given that worker shortages exist, and terminating an unproductive employee will tend to make that challenge even greater.

It must be recognized that one improperly behaving team member can make it difficult for many workers to enjoy their jobs. Taking positive action to retain employees as a group is always more useful than retaining a single individual who may continually be disruptive to a team.

Internal Recruiting

For many foodservice operators, candidate recruitment is an ongoing activity. One good way to address the team member search process is to consider it to be either an internal or external process. An **internal search** is used when the best candidates for vacant positions are believed to be currently employed by the operation.

Key Term

Internal search (employee): A promote-from-within recruitment approach used to identify qualified job applicants.

Done properly, an internal search approach can be effective. Current employees may inform others about pending job openings in individual conversations, in employee meetings, and by posting information on employee bulletin boards, websites, newsletters, or other media. Advantages of using internal searches include that they

✓ build employee morale;
✓ can be initiated quickly;
✓ improve the probability of making a good selection because much is already known about the selected individual;
✓ are less costly and time consuming than initiating external searches;
✓ result in reduced training time and fewer training costs because the individual selected need not learn about organizational topics with which he/she is already familiar;
✓ encourage talented individuals to stay with the organization.

Key Term

Employee referral: A recruitment approach that occurs when a current foodservice employee refers someone they know to fill an operation's vacant position. Employee referrals are often seen as a good way to find new employees because the referred employee is already known to someone who works in the operation.

While not technically an internal search system, some operators utilize **employee referral** systems to discover potential applicants recommended by current employees.

Employee referral systems often work well because employees rarely recommend someone unless they believe the persons recommended can do a good job and fit well in the organization. Also, existing employees often have an accurate view of the job including the organization's culture. This information reduces unrealistic expectations and can also help reduce new employee turnover. In some operations, a financial bonus is paid to a team member who recommends a person that is hired by and remains with the organization a specified time.

There are potential challenges with employee referral systems including that those employees who recommend one or more persons may suggest their friends or relatives regardless of those individual's qualifications. For this reason, the same standards of employment consideration that apply to other individuals considered for a position should be used for all referred candidates.

Traditional External Recruitment Methods

Even when internal recruiting methods are effectively used, foodservice operators must often conduct an **external search** outside their organizations for new job applicants.

Commonly used strategies for external searches include:

Key Term

External search (employee): An approach to seeking job applicants that focuses on candidates not currently employed by the organization.

✓ *Traditional advertisements*—These can range from simple "Help Wanted" or "Now Hiring" postings on-site or on public bulletin boards. Some ad venues are free or low cost. These include social media sites and bulletin boards at apartment complexes, childcare centers, supermarkets, libraries, community centers, and school newspapers. Employers seeking candidates for whom English is not their primary language may also target foreign-language newspapers and newsletters.

✓ *Educational institutions*—Most educational institutions provide services to help their graduates find jobs. Whether the new job requires a high-school diploma, specific vocational training, or a degree, educational institutions can be useful sources of qualified job candidates. Technical schools and secondary (high) schools in an employer's area typically offer employers the chance, at little or no cost, to assess the quality of their students.

✓ *Unsolicited applications*—Foodservice operators often receive unsolicited applications or requests to be considered for job openings. These may arrive in an operator's office by letter, text, or e-mail, or in-person delivery. Even when there are no current vacancies, these applications should be initially reviewed and then be kept on file.

Note: Unsolicited applications submitted to a foodservice operator have a relatively short life span because those searching for jobs typically continue their search until it ends successfully. Many operators review unsolicited applications submitted daily, and some even arrange on-the-spot interviews when good candidates are scarce.

Internet-based External Recruitment Methods

The Internet has significantly changed how foodservice operators recruit employees. Increasingly, popular social media sites are used to communicate with applicants about vacancies, questions, and other information. Also, many foodservice operators feature general "Career Opportunity" or "Current Openings" information on their websites along with more specific information about current position vacancies.

Career-related information on well-developed company websites may include:

✓ Organization overview including press releases, current news stories, and awards
✓ Corporate culture information about the organization's values, vision, and mission
✓ General information about the organization's history, operating statistics, and employee testimonials
✓ Current position vacancies
✓ Compensation ranges including a description of benefits

Websites may also include application forms and e-mail or other addresses for additional information. This is particularly useful for attracting younger workers who prefer online applications. Many hospitality organizations utilize social media sites such as Instagram, ZipRecruiter, LinkedIn, and Monster. As well, their own Facebook or other social media site pages can offer information about employment opportunities. X (formerly Twitter) is another popular site that allows foodservice operators to announce position vacancies and receive inquiries from those who are potentially interested in the positions. For management-level openings, Hcareers is a site that provides a huge database of hospitality industry–related employment opportunities.

Technology at Work

Social media sites that emphasize and feature job openings are increasingly popular and especially so with younger workers. The popularity of these sites, however, can change very rapidly. As a result, foodservice operators must make it a point to stay up to date with changes in those social media sites that reflect increasing or decreasing popularity.

(Continued)

Monitoring changes in such sites certainly need not be a daily activity, but it should be at least a monthly or quarterly activity. There is little doubt that the popularity of specific social media sites and apps will continue to increase and decrease as they have in the past.

To learn more about the most currently popular social media sites used by job seekers, enter "Popular social media employment sites" in your favorite search engine and view the results.

Choosing Team Members

The proper selection of new team members is an important task. Numerous advantages accrue to operators when they select new employees properly. This is so because effective employee selection:

Saves time: When the right employee is hired the first time, it saves the time and effort operators must otherwise spend to recruit and interview. Newly-employed workers with the proper characteristics can more quickly and effectively understand their new roles and responsibilities, and less time is spent on training.

Decreases hiring costs: There are real costs to hiring new workers. Those costs may include advertising vacancies and various administrative costs of hiring. Choosing the right new employee (the first time!) helps minimize hiring costs.

Reduces burn out of existing employees: There are numerous times when foodservice operators hire new employees because they are immediately needed. When the right employee is hired, that new team member can quickly contribute to the team. Then existing staff might have to work fewer overtime hours. Also, more experienced staff can more easily handle excessive workloads while new staff members learn about required tasks.

Enhances morale: When the "right" new employees are hired, they can quickly adjust and contribute their skills and experience, and they can also encourage other employees in their own jobs. Alternatively, a bad hire can have the opposite effect. Poor hires, including those with negative attitudes and unsatisfactory work ethics, can detract from an operation's culture and potentially frustrate current employees.

After foodservice operators recruit and identify a pool of qualified candidates, they must select the best applicants to fill vacant positions. Several employee selection tools can be used including:

✓ Applications
✓ Interviews
✓ Pre-employment Testing
✓ Background Checks and References

Applications

An employment **application** should be completed by all employment applicants, and effective application forms do not need to be extremely lengthy or complex.

The employment application helps foodservice operators learn information to determine whether the applicant can, with appropriate training, perform a job's essential functions. The requirements for a legitimate and legally sound application should focus exclusively on job qual-

Key Term

Application: A form, questionnaire, or similar document that employment applicants must complete for the employer. An application may exist in a hard copy, electronic copy, or Internet medium.

ifications. Foodservice operators should have proposed employment applications reviewed by an attorney who specializes in employment law before they are used.

A foodservice operator generally uses the same application for all job applicants. Doing so helps create and ensure a structured and consistent assessment process. It most often begins with an application review, and then proceeds through one or more rounds of interviews. The use of uniform applications enables easy comparisons of one applicant to another in a fair and consistent manner. A well-constructed application should also have a signature section that allows applicants to sign the document indicating all information they have provided is true and accurate to the best of their knowledge.

Technology at Work

To compete effectively in an increasingly tight labor market, foodservice operators should make it easy for potential employees to apply for a job. One good way to do this is to ensure that an operation has an easy-to-use online job application program.

An online job application is a form on the Internet where applicants can communicate their skills and relevant experience for a specific job or position. Foodservice operators use online applications to help speed up the hiring process and increase the number of potential job candidates for the position.

Several technology companies offer online application services appropriate for small businesses. To learn more about them and how they can be publicized to potential job applicants, enter "Online job application programs for restaurants" in your favorite search engine and view the results.

Figure 8.1 is an example of an employment application that is legally sound. Note specifically how questions relate only to the ability to work, work history, and job qualifications.

Lawson's Restaurant

Application for Employment

Name: _____ , _____ _____
         Last   First   Middle

Street Address _____ , _____ _____
                        City   State   Zip

Telephone (_____) _____-_____   Social Security # _____-_____-_____

Position Applied for _____

How did you hear about this opening? _____

When can you start? _____/_____/_____

Desired Hourly Wage: $_____

                                                                                                    Yes   No

- Are you a U.S. Citizen or otherwise authorized to work in the U.S.?   ☐   ☐
- Are you capable of performing the essential functions of the job
  you are applying for with or without reasonable accommodation?   ☐   ☐
- If applying for a job involving the service of alcoholic beverages,
  are you over the age of 21?   ☐   ☐
- Are you under the age of 18? Birth Date: _____/_____/_____   ☐   ☐
- Are you looking for full-time employment?   ☐   ☐
- If no, what days and hours are you available? (please list all that apply)

	From	To
Monday		
Tuesday		
Wednesday		
Thursday		
Friday		
Saturday		
Sunday		

- Do you have dependable means of transportation to and from work?   ☐   ☐
- Do you have any criminal charges pending against you?   ☐   ☐
- Have you been convicted of a felony in the past seven years?   ☐   ☐

   If yes, please fully describe the charges and disposition of the case:

   _____

   _____

   (Note: Conviction of a felony will not necessarily disqualify you from employment)

Education:

	School Name/ Location	Year Completed	Major/ Degree
High School:			
Technical School:			
College:			
Other:			

Figure 8.1   Sample Employment Application

In addition to your work history, are there other certifications, skills, and qualifications we should know about? _____

_____

Employment History. (start with the most recent employer)

Company Name: _____

Location: _____

Starting Position: _____

Start Date: _____/_____/_____

Ending Position: _____

End Date: _____/_____/_____

Name of Supervisor: _____

                May we contact: Yes _____ No. _____

Responsibilities: _____

_____

Reason for Leaving: _____

_____

Company Name: _____

Location: _____

Starting Position: _____

Start Date: _____/_____/_____

Ending Position: _____

End Date: _____/_____/_____

Name of Supervisor: _____

                May we contact: Yes _____ No. _____

Responsibilities: _____

_____

Reason for Leaving: _____

_____

I state that the facts written on this application are true and complete to the best of my knowledge. I understand that if I am employed, false statements on this application can be considered cause for dismissal. The company is hereby authorized to make any investigations of my prior educational and employment history. I understand that employment at this company is "at will" which means that I or the company can terminate the employment relationship at any time, with or without prior notice. I understand that no supervisor, manager, or executive of this company, other than its owner has the authority to alter the at will status of my employment.

I authorize you to make such legal investigations and inquiries into my personal and employment, criminal history, driving record, and other job-related matters as may be necessary in determining an employment decision.

_____   _____

Signature   Date

Figure 8.1 *(Continued)*

Interviews

Some job candidates who have submitted employment applications will be selected for one or more interviews. These can be powerful selection tools; however, the types of questions that can legally be asked in an interview are restricted. If interviews are improperly performed, significant legal liability can result. I If an applicant is not hired based on his/her answer to, or refusal to answer, an inappropriate question, that applicant may have the right to file (and perhaps win!) a lawsuit.

The **Equal Employment Opportunity Commission (EEOC)** suggests an employer consider three issues when deciding whether to include a particular question on an employment application or in a job interview:

Key Term

Equal Employment Opportunity Commission (EEOC): The entity within the federal government assigned to enforce the provisions of Title VII of the Civil Rights Act of 1964.

1) Does the question tend to screen out minorities or women?
2) Is the answer needed to judge this individual's competence for job performance?
3) Are there alternative, nondiscriminatory ways to judge the person's qualifications?

Note: The EEOC enforces the antidiscrimination provisions of Title VII. The EEOC investigates, mediates, and sometimes even files lawsuits on behalf of employees. Businesses that are found to have discriminated against one or more employees can be ordered to compensate the employee(s) for damages in the form of lost wages, attorney fees, and other expenses.

Foodservice operators, or those designated to make hiring decisions, must carefully select interview questions. In all cases it is important to remember the job itself dictates whether questions are allowable. The questions asked of all applicants should be written down in advance and carefully addressed. In addition, supervisors, co-workers, and others who may participate in the interview process should be trained to avoid asking inappropriate questions that could increase an operation's potential liability.

In general, age is considered irrelevant in most hiring decisions, and questions about date of birth are improper. However, age can be a sensitive pre-employment question because the Age Discrimination in Employment Act protects employees 40 years old and above. Asking applicants to state their age if they are younger than 18 years old is permissible because in most young persons are permitted to work only a limited number of hours each week. It may also be important when hiring bartenders and other servers of alcohol to confirm that their ages are at or above the state's minimum age for serving alcohol. Questions about race, religion, and national origin are always inappropriate, as is the practice of requiring applicant photographs that must be submitted before or after an interview.

As they conduct their employment interviews, foodservice operators must recognize that safe questions can be asked about an applicant's present employment, former employment, and job references. Questions asked on the application and during the interview should focus on the applicant's job skills and nothing more.

What Would You Do? 8.1

I liked her personality a lot, but I just didn't like her looks," said Margie, the cafeteria supervisor.

"What didn't you like?" asked Terri, the head Dietitian for the hospital foodservice where Margie worked and the hiring manager for the hospital's foodservice operation.

Margie and Terri were discussing Aurora, a 20-something young woman who had just finished an interview for a job bussing tables in the hospital's "Open to the public" dining room.

"Well, I don't know if its "Goth," or "Emo," or whatever kids call it today, but all that black around her eyes and on her lips, and the multiple ear piercings. All those hoops! It just seems odd to me," said Margie.

"I agree it looks different," said Teri, "but I remember when I was about her age my parents thought the way I dressed was outlandish! Now I guess they would call the outfits I wore back then old school!"

"Well, she wouldn't be in violation of our no visible tattoo policy," said Margie, "because she didn't have any facial tattoos. But maybe we need a better policy on appropriate dress?"

"Well, we certainly have the legal right to establish an appropriate dress code," said Terri, "but if we get into clothing colors, earrings, eye make-up styles, and lipstick colors, I think we might start to be on a slippery slope with our other employees!"

Assume you were Terri. Employers are free to regulate the physical appearance of their employees when it is done in a nondiscriminatory and equitable way. They can also refuse to hire employees that do not conform to standards addressed in company policy. Assuming she was qualified, would you be in favor of hiring Aurora? Do you agree with Margie that you need a policy that specifically addresses clothing and make-up colors? Explain your answer.

Pre-employment Testing

Pre-employment testing improves the employee screening process. The reason: test results can measure the relative strengths of two applicants. In the foodservice industry, pre-employment testing is generally of three types: skills, psychological, and drug screening.

Skills tests include activities such as typing tests for office workers, computer application tests for those involved in managing websites, and food production tasks for chefs and cooks. Psychological testing can include personality and other tests that predict performance or mental ability.

Pre-employment drug testing is allowable in most states, and it can reduce insurance rates and potential worker liability issues. Many foodservice operators believe a drug-free environment attracts better applicants with the resulting effect of a higher quality workforce. When pre-employment drug testing is used, care is needed to ensure the accuracy of results. (Sometimes applicants with erroneous test results have lost their job offers and successfully sued the employer.)

The laws surrounding mandatory drug testing are complex. Foodservice operators who elect to implement a voluntary or mandatory pre- or post-employment drug testing program should first seek advice from an attorney who specializes in labor employment law in the jurisdiction where the operator does business.

Background Checks and References

Foodservice operators sometimes use background checks before hiring workers for selected positions since many resumes and employment applications are falsified. While many types of background checks are available, not all are advisable. Background checks should be specifically tailored to obtain only information relating directly to each applicant's employment suitability.

Commonly used background checks include:

Criminal history—As a general rule, criminal conviction records should be checked when there is a possibility that the person could create significant safety or security risks for co-workers, guests, or others. Examples include employees who will have close contact with minors, the elderly, people with disabilities, and medical patients, and those with access to weapons, drugs, chemicals, or other potentially dangerous materials.

Driving records—Motor vehicle records (MVRs) are available from state motor vehicle departments. They usually contain information about traffic violations, license status, and expiration dates. MVRs should be checked for any employee who will drive a company or personal vehicle for the employer's business including making food deliveries to guests.

Academic credentials and licenses—While less often addressed, academic information such as schools attended, degrees awarded, and transcripts should be verified when a specified level or type of education is necessary for a particular job. Similarly, proof of licenses and their current status, expiration dates, and any past or pending disciplinary actions should be obtained if a license is required for the position in question.

Foodservice operators should always obtain written consent before conducting any background check. This helps protect against invasion of privacy, defamation, and other wrongful act claims. It is also useful to expand the waiver language on consent forms to include the employer and those who assist with background checks including staff, former employers, and screening firms.

Key Term

Negligent hiring: Failure of an employer to exercise reasonable care when selecting employees.

Key Term

Negligent retention: Retaining an employee after the employer becomes aware of an employee's unsuitability for a job by failing to act on that knowledge.

Failure to conduct background checks on applicants for some positions can subject operators to legal difficulty under the doctrines of **negligent hiring** and **negligent retention**.

Negligent hiring liability usually occurs when an employee (a) has caused injury or harm and had a reputation or record that showed the propensity to do so, and (b) the record would have been easily discoverable if reasonable care (a diligent search) had been shown. Similarly, negligent retention may be charged if an employer hires an employee and then discovers disqualifying information but does not remove the employee from the job.

Required Documentation for New Team Members

After an employment offer has been made and accepted, foodservice operators have one final task to complete before an applicant can become an "official" team member. That task involves obtaining the proper documentation for the new worker.

Required documentation for newly chosen team members include:

✓ Form I-9 (Federal Form)
✓ Form W-4 (Federal Form)
✓ State and Local Tax Forms

Form I-9 (Federal Form)

All newly hired employees are required to fill out an Employment Eligibility Verification form (commonly known as an I-9 form) stating that they are authorized to work in the United States. U.S. Citizenship and Immigration Service (USCIS) regulations allow an individual 72 hours from time of hire in which to complete Form I-9.

The Form I-9 requires potential employees to verify their citizenship status and legal eligibility to work. There are numerous documents that an employee can use to provide this information including a U.S. passport, driver's license, Social Security card, and school identification card.

Under current law, employers are not required to verify the authenticity of the identification documents they were presented, but they must keep a copy of them on file. The documents must pass a good-faith test ("Do they look real?") and, if they do, the applicant may be hired.

If an employee is later found to be unauthorized, the employer must terminate the worker's employment. Employers who do not end employment of unauthorized workers or who knowingly hire unauthorized workers can face significant fines.

Find Out More

E-Verify, authorized by the Illegal Immigration Reform and Immigrant Responsibility Act of 1996 (IIRIRA), is a web-based system through which employers electronically confirm the employment eligibility of new employees. In the E-Verify process, employers create "cases" (individual employee records) based on information taken from an employee's Form I-9, Employment Eligibility Verification.

E-Verify then electronically compares that information to records available to the U.S. Department of Homeland Security (DHS) and the Social Security Administration (SSA). The employer usually receives a response within a few seconds that either confirms the employee's employment eligibility or indicates the employee needs to take further action to complete the case.

E-Verify is administered by Social Security Administration and U.S. Citizenship and Immigration Services (USCIS). USCIS facilitates compliance with U.S. immigration law by providing E-Verify program and user support, training, and outreach, and developing innovative technological solutions in employment eligibility verification.

The use of E-Verify is not mandatory (except in certain states), but its use is always a good idea. To find out more about the E-Verify program and how it helps employers stay in compliance with the law, enter "Why use E-Verify?" in your favorite search engine and review the results.

Form W-4 (Federal Form)

Foodservice operators hiring new employees must also ensure the employee completes a Federal Form W-4. This form (formally titled "Employee's Withholding Certificate,") is an IRS form that employees fill out and submit to their employers when they begin a new job.

Essentially, employers use the information provided on a W-4 to calculate how much tax to withhold from an employee's paycheck throughout the year. Form W-4 tells a foodservice operator (the employer), about the employee's filing status, multiple jobs adjustments, and amount of allowable tax credits. Other information provided includes the amount of other income, amount of deductions, and any additional amount to withhold from each paycheck. These are important because the data is used to compute the amount of federal income tax to deduct and withhold from the employee's pay. If an employee fails to submit a properly completed Form W-4, the employer must withhold federal income taxes from their wages as if they were single or married filing separately.

State and Local Tax Forms

In addition to federal income taxes, workers in many states must pay various state (and sometimes local) income taxes as well. Currently, eight states in the United States choose not to impose an income tax. In the majority of states, however, income taxes are assessed in a manner similar to the federal government's approach. But slight differences do exist in some states and, when they do, the state may require employers to also utilize a separate state form that is typically similar to the Federal Form W-4.

Training Team Members

Foodservice operations are labor-intensive. Technology has significantly affected how foodservice operations use the internet for marketing, and software advances allow foodservice operators to collect and analyze guest data efficiently, however, technology generally has not affected the number of employees required to produce and serve the menu items foodservice guests want to buy.

Whether the operation is a food truck or a high-end table service restaurant, newly hired team members must acquire the knowledge and skills needed to excel. Also, more experienced peers must often obtain new knowledge and skills to keep up with changes in work procedures, menu items, and services offered to guests. As a result, proper **training** is critical for both an operation's new and existing team members.

> **Key Term**
>
> **Training:** The process of developing a staff member's knowledge, skills, and attitudes necessary to perform required tasks.

Benefits of Training

Several benefits arise when its operator makes staff training a priority:

1) *Improved worker performance*—Trainees learn knowledge and skills to perform required tasks more effectively, and their on-job performance improves.
2) *Reduced operating costs*—Improved job performance helps reduce errors and rework, and associated costs are reduced. Workers performing their jobs correctly will be more productive with fewer labor hours needed.

3) *More satisfied guests*—Proper training can yield more service-oriented employees.

4) *Fewer operating problems*—Busy operators can focus on priority concerns rather than routine operating problems caused by inadequate training.

5) Lower employee turnover rates—Fewer new staff members become necessary as employee turnover decreases. Properly trained employees rewarded for successful performance are less likely to leave, and operators have less need to recruit new employees.

6) *Higher levels of work quality*—Effective training identifies quality standards that define acceptable product and service outputs. Trained employees are more interested in operating equipment correctly, preparing menu items the "right" way, and properly interacting with guests.

7) *Improved morale*—Staff morale often improves as employees recognize their employer's training commitment that allows advancement to more responsible and higher-paying positions. High levels of morale, in turn, are important to the success of the training process and the operation.

8) *Easier to recruit new staff*—Satisfied staff tell their family and friends about the positive work experiences, and their contacts may become candidates for future position vacancies. Foodservice operations that emphasize training can evolve into "employers of choice" providing "first choice" rather than "last chance" employment opportunities.

9) *More professional staff*—Professionals want to do their job as best they can, and this is only possible with appropriate training.

10) *Greater profits*—If guests are more satisfied and revenues increase, and if labor and other operating costs are reduced, there is significant potential for increased profits. Training must "add value" measured by the difference between increased profits and added training costs. While this measurement is not easy, most industry observers believe that, if done correctly, training does not "cost," but rather, it "pays!"

Find Out More

Those with training experience recognize that they often encounter generational differences in the trainees' learning styles. A generation is a group of people born during a particular time. Individuals in these generations often operate with similar attitudes, ideas, and values that might impact their learning styles.

While examples of generational differences abound, one easy way to recognize situations directly affecting employee training involves familiarity with the use of technology. For example, the "Baby Boomer" generation (born between 1946 and 1964) currently make up approximately 25% of the workforce. Millennials (Generation Y) (born between 1981 and 2000) currently make up approximately 35% of the workforce. In many cases, the exposure of the Baby Boomer generation to advanced technology is often less useful. Those in this generation still recall when answering the phone meant going to a physical location and picking up a receiver.

Millennials, however, grew up with advanced technology and are comfortable with its use in many settings. Technology may be one area where trainers need to consider generational differences as the delivery of training programs evolve.

To learn more about other areas in which generational differences can impact training, enter "Generational differences in the workplace" in your favorite search engine and review the results.

Characteristics of Successful Trainers

Every foodservice operator will experience the need to train new team members. Sometimes foodservice operators can provide the training, but not always. This is especially so in a large foodservice operation when others must provide new employee training.

Who should do training? Sometimes this question is answered by asking other questions: Who is available? Who wants to do it? Who has the time? Who will complain least if given the assignment? Who is a good "people person" that can interact well with the employee needing training?

While these factors are often relevant, others may be more important, including the following characteristics required for successful trainers:

✓ *Have the desire and resources to train*—Successful trainers want to train and they like helping others. There is often recognition for a job well done, and they know effective trainers are frequently promoted to higher level positions. However, foodservice operators know that stress resulting from inadequate training resources is often a disincentive for accepting and successfully completing a training assignment.

✓ *Have the proper attitude about the employer, peers, position, and the training assignment*—Foodservice operators who emphasize the importance of staff members and provide quality training opportunities to employees at all levels increase the morale of their trainers. Conversely, when training is just another and not so important responsibility, a less-than-willing attitude toward being a trainer can result.

✓ *Possess the necessary knowledge and skills to do the job for which training is needed*—Effective trainers must be knowledgeable about and have the skills necessary to perform the work tasks for which they will train.

✓ *Use effective communication skills*—Trainers are effective communicators when they (1) speak a language the trainee understands, (2) recognize body language is a powerful communication method, (3) use a questioning process to learn what the trainee learned, and (4) speak to effectively communicate, not merely to impress.

✓ *Know how to train*—The importance of **train-the-trainer programs** to provide trainers with skills to train effectively should be obvious but often is overlooked.

> **Key Term**
>
> **Train-the-trainer (program):**
> A training framework that turns employees into subject matter experts who can teach others.

✓ *Have patience*—Few trainees learn everything they must know or be able to do during initial training. Effective trainers understand that training must sometimes be repeated several times and differently. They know the goal is not to complete training quickly; rather, it is to provide the knowledge and skills the trainees need.

✓ *Have time to train*—Effective training takes time and must be scheduled for the trainer and the trainees.

✓ *Show genuine respect for the trainees*—Trainees must be treated as professionals. Trainers know that those whom they respect will also respect them, and mutual respect allows training to be more effective.

Steps In Creating Effective Training Programs

The creation of effective training programs is an ongoing activity. As shown in Figure 8.2, to plan effective training operators must use a specific five-step process that requires them to:

Step 1: Identify Training Needs
Step 2: Develop Training Objectives
Step 3: Develop Training Plans
Step 4: Create Training Lessons
Step 5: Develop (up-date) Training Handbooks

Each step in the planning process is important. Understanding and completing each step is essential if a foodservice operation is to develop high quality training programs.

Step 1: Identify Training Needs

Every foodservice operation has training needs, but they vary depending on the operation itself. There are many examples of short-term operating and other concerns that can be addressed by training. New work methods, new equipment or technology, and/or implementation of cost-reduction processes may be helpful and assist with focused training. As well, there can be a need for more or different products and services to meet guests' evolving needs.

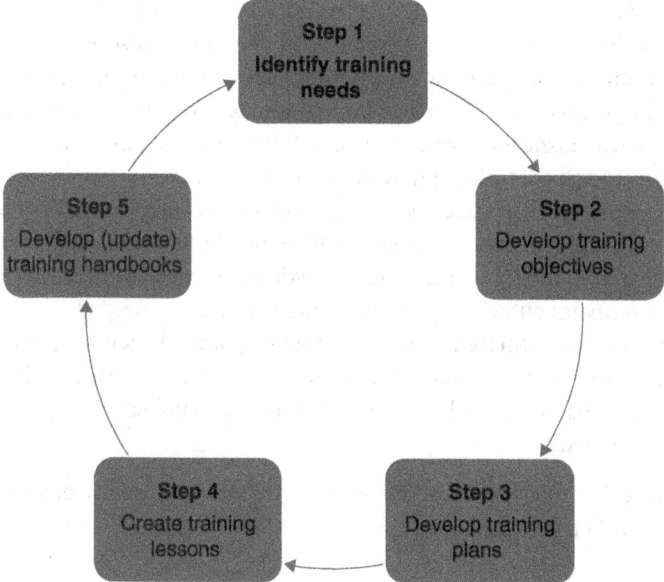

Figure 8.2 Steps in Creating Effective Training Programs

To be cost-effective and successful, a training program must provide time saving and economic benefits that outweigh its costs. The success of training can then be demonstrated by considering how worker knowledge, skills, and performance improve because of the training.

Operators can determine their training needs in several ways.

✓ *Listening to guests*—Training programs have been developed to enable guests to receive what they want, and an objective drives the training to perform this task. For example, there may be a preferred way to make a dinner reservation for a guest at 8:00 p.m. For example, assume a person in line for a food truck meal orders French fries "without salt." Since this is a common request by guests, the cook that prepares the fries must be taught the correct way to prepare French fries as a salt-free menu item.

✓ *Listening to current staff members*—Some foodservice operators use suggestion boxes, open-door policies, and/or frank input from performance appraisals and coaching sessions to identify problems resolvable with proper training.

✓ *Observing current work performance*—Those who "manage by walking around" may note work procedures deviating from an established **standard operating procedure (SOP)**. It is important that proper performance is clearly defined so employees know what is expected of them, and foodservice operators know when performance is acceptable.

Key Term

Standard operating procedure (SOP): The term used to describe how something should be done under normal business conditions.

✓ *Reviewing inspection scores*—All foodservice operations are subject to inspections by local health departments (see Chapter 1). In addition, foodservice operators may create internal inspections for safety-related conditions and proper opening or closing procedures. Franchised operations are regularly inspected by franchisors. When inspection scores are carefully reviewed, there is help identifying where additional training may improve the scores.

✓ *Monitoring online reviews*—Successful foodservice operators regularly monitor the scores and comments they receive on websites allowing guests to describe their interactions with an operation. If guests regularly indicate a specific situation, the stated problems may point out where new or additional training is needed.

✓ *Analyzing financial data*—Differences between budget plans and actual operating data may suggest negative variances traceable to problems with training implications. After problems are identified, corrective actions including training may be implemented.

The methods in which foodservice operators identify their training needs are varied, and as many different methods as possible should be utilized.

Step 2: Develop Training Objectives

Training objectives specify what trainees should know and be able to do when they have successfully completed their training. Those planning training programs must know what the training should accomplish, and training objectives help them do this.

Training objectives are critical to training evaluation, and they should describe the expected training results rather than the training process itself. For example, assume a foodservice operator has identified the need to train servers to properly use the operation's point of sale (POS) system (See Chapter 2). In this situation, the proper training objective is to ensure all trainees can properly use the POS system.

One good way to develop training objectives is to complete a sentence similar to the following for each training need identified in Step 1:

> As a result of the satisfactory completion of the POS training session, the trainee will . . .

Training objectives must be reasonable (attainable), and they must be measurable. Objectives are not reasonable when they are too difficult or too easy to attain. The concern that training objectives be measurable relates to the role of the objectives in the training evaluation process.

For example, how can training programs be evaluated if success is measured by objectives such as those listed below?

✓ Trainees will *realize* the importance of proper POS system usage.
✓ Trainees will *understand* how the POS system works.
✓ Trainees will *recognize* the need to use the POS system properly.

Contrast the above with measurable objectives for the same topics:

✓ Trainees will *demonstrate* they can log their own server identification information into the POS system.

✓ Trainees will *show* their trainer that they can properly enter guests' orders into the POS system.

✓ Trainees will *confirm* to the trainer that they can properly produce a guest check total suitable for presentation at the end of a guest's meal.

The best training objectives typically use an action verb to tell what the trainee must demonstrate or apply after training. Examples include *operate, calculate, explain, show, produce,* and *assemble.* By contrast, unacceptable verbs that cannot be readily measured include *know, appreciate, believe,* and *understand.*

Step 3: Develop Training Plans

Training plans organize training content and provide an overview of the structure and sequence of the training program. They show how individual training lessons should be sequenced to teach required knowledge and skills.

Suggestions for determining the sequence for subject matter in a training plan include:

Key Term

Training plan: A description of the overview and sequence of a complete training program.

✓ Beginning with an introduction to tell why the training is important and how trainees will benefit from it.

✓ Providing an overview of training content.

✓ Planning training lessons to progress from simple to complex. Simple information presented at the beginning of a training session allows trainees to quickly feel comfortable in the learning situation. It also provides the confidence needed to master the program.

✓ Building on the trainees' experiences. Combine unfamiliar information with familiar content to allow trainees to build on their experience.

✓ Presenting basic information before more detailed concepts are discussed.

✓ Progressing from general information to specific information. For example, provide an overview of how to set up a dining room and then begin with specific steps.

✓ Considering the necessity of "nice-to-know" and "need-to-know" information. Basics should be presented before other information and addressing "why" before the "how" is generally best.

✓ Using a logical order and clearly identify what information is prerequisite to other information.

Training plans allow trainers to (1) plan the dates and times for each training lesson, (2) consider the topic (lesson number and subject), (3) state the training location, (4) indicate the trainer(s) who conduct the training, and (5) determine the trainees for whom specific training lessons are applicable.

Step 4: Create Training Lessons

A **training lesson** provides the information needed to present a single session that is part of a broader training plan. In effect, it is a "turnkey" module that defines a specific training session:

✓ *Why*—the objective(s) of the training session
✓ *What*—the content of the training session
✓ *How*—the method(s) used to present the training

Key Term

Training lesson: Information about a single session of a training plan. It contains one or more training objectives and indicates the content and method(s) to enable trainees to master the content.

Individual training lessons may be as short as a few minutes or as long as several hours. A training lesson can teach new staff members how to perform a single task (e.g. operating an oven) or it can be used to teach experienced staff new steps in a single task. Training lessons may be designed for one trainee at a time and/or to trainee groups.

Step 5: Develop (and update) Training Handbooks

Designing effective training programs requires thought, time, and creativity. The process is most cost-effective when training plans, lessons, and resource materials developed are used for more than one training experience. For that reason, experienced foodservice operators know they must develop (and keep updated) an easy-to-utilize **training handbook** specific to their operations.

A wise foodservice operator maintains training information in an organized fashion to allow easy replication of training and makes handbook revisions, as necessary. This benefits operators since time and money spent to develop training tools need not be replicated.

Key Term

Training handbook: A hard copy or electronic manual containing the training plan and associated training lessons for a foodservice operation's complete training program.

Evaluation of Training Efforts

Evaluation is the final step in the employee training process, and it indicates if the training was successful (were goals attained?) or unsuccessful (goals were not attained).

Foodservice operators' time, money, and labor are increasingly limited. Operators must determine if their commitment of resources to planning and implementing training procedures is a better use than other alternatives. This is why training evaluation is important. Additional reasons to evaluate training efforts include to:

✓ *Assess whether training achieved planned results*
✓ *Identify strengths and weaknesses of training*
✓ *Determine success of individual trainees*
✓ *Determine trainees' need for future training*
✓ *Assess costs and benefits of training*

Some trainers think about training evaluation only in the context of an after-training assessment. While training should be evaluated at its completion (and perhaps again several months after its completion), evaluation can also be helpful before training even begins, and while it is conducted.

When evaluation methods have the above characteristics, they may be used to evaluate training before, during, and/or after training is delivered.

Evaluation Before Training

To understand evaluation before training, assume that a trainee participated in a food safety training session and, after the training was completed, a written test was administered. Assume further that the trainee missed only two questions of the 20 that were asked. Many trainers would likely conclude that the training was successful because the participant scored 90% (18 questions correct ÷ 20 questions total = 90%).

In fact, the training may have wasted the operation's resources and the trainee's time if the trainee *already* knew the information taught before the session began. In this example, the after-training evaluation really measured what the trainee knew when the training was completed, rather than what was learned as a direct result of the training session.

To address this concern, some trainers use a **pretest/post-test evaluation** tool in which key concepts to be addressed during the training are identified, and these concepts are addressed in a pretest administered before the training begins.

Then the same measurement tool using the same questions is administered at the end of training. The improvement (change) in scores between the pre- and post-tests represents one measure of training effectiveness.

Key Term

Pretest/post-test evaluation: A before and after assessment used to measure whether the expected changes took place in a trainee's knowledge, skill level, and/or attitude after the completion of training.

Evaluation During Training

Some training evaluation can occur during the training session. For example, trainers can use an introductory session statement to state they will ask for feedback during the session. When this feedback is solicited, trainers can obtain a reality check and perhaps learn helpful information to improve the remainder of the training.

Trainers facilitating group sessions can ask trainees to write anonymous responses to statements such as:

✓ I wish you would stop doing (saying). . .
✓ I hope we continue to. . .
✓ I don't understand. . .
✓ I hope you will begin to. . .
✓ A concept I wish you would discuss further is. . .
✓ A concept I want to learn more about not yet discussed is. . .

The major goal of "during the training" evaluation is to learn how to maximize use of the remaining training time. Then revisions to training content and/or delivery methods can better ensure attainment of training objectives.

Evaluation After Training

After-training evaluation can assess the extent to which training achieved its planned results, and it may also identify how training sessions might be improved. Several after-training evaluation methods are in common use. One or more of these methods can provide information for foodservice operators to improve their future training efforts.

Common after-training assessment methods include:

✓ *Observation of after-training performance*—Owners, operators, and supervisors can "manage by walking around" and observe whether knowledge and skills taught during training are being applied. For example, storeroom personnel can be observed as they receive incoming products, and procedures used can be compared with those presented during training. Note that, when proper procedures are used, a "Great job!" compliment is in order. By contrast, a coaching activity to remind staff members about incorrectly performed procedures may also be needed.

✓ *Objective tests*—These can be written, oral, and/or skill-based and include traditional written exams and quizzes or after-training demonstrations. Written examinations including multiple choice and true/false questions are most often used in foodservice operations because they are **objective tests**.

Key Term

Objective test: An assessment tool such as a multiple choice or true/false test whose questions have one correct answer and yield a reduced need for trainers to interpret trainees' responses.

In an objective test with only one correct answer, little or no interpretation is needed, and minimal time is required for trainees to complete the exam and for trainers to score it. Analysis of incorrect answers can often lead trainers to identify areas in which training emphasis were insufficient, and/or trainee understanding was deficient.

✓ *Third-party opinions*—Feedback from guests can help assess training programs that addressed aspects of products and services that affected them. The use of a **mystery shopper** in some foodservice operations is another example. Feedback can also be generated by comment cards, interviews, and/or follow-up surveys with guests in person or online.

✓ *Analysis of user-generated content (UGC) site scores*—UGC scores (see Chapter 5) are increasingly critical in the successful marketing and operation of a profitable foodservice business. Just as UGC site scores are reviewed to indicate potential areas where training is needed, training lessons developed to address the issues identified on UGC sites should yield increased scores.

> **Key Term**
>
> **Mystery shopper:** A person posing as a foodservice guest who observes and experiences an operation's products and services during a visit and who then reports findings to the operation's owner. Also referred to as a "secret shopper."

✓ *Analysis of operating data*—In some cases, an assessment of operating data can yield valuable information about the effectiveness of training efforts. For example, training that addresses guest service and food costs should yield, respectively, increased guest service scores and lowered food costs (if components of these data can be separated). Then the separate parts can determine how they were influenced by training. Training designed to help servers sell specific entrees or desserts, for example, could be measured by examining pre- and post-training sales data for these items.

Regardless of the metric(s) used to assess an operation's training efforts, post-training evaluation is a key part of an operation's overall training effort, and it should not be neglected.

With a carefully selected and properly trained staff, foodservice operators are in a good position to produce and serve products. This, in turn, will ensure that guests return again. The control of production and service is essential to a foodservice operation's success, and doing this effectively is the topic of the next chapter.

What Would You Do? 8.2

"Seriously, you want me to train a new server again?" said June Hernandez, a long-time server at the Pink Flamingo Seafood restaurant.

"You're our best server, and you know it, June. I just want to make sure that our new hire gets the best possible chance of success, and he will if you do the training," replied Demonte Jackson, the dining room supervisor at the Pink Flamingo.

Demonte had just hired a new server to replace one that had recently resigned from the restaurant. They were discussing the newly hired server's initial training, and Demonte wanted June to assume that responsibility.

"Here's the thing Demonte," said June, "when I train a new server, I get less tables because of the time it takes to show them the ropes. Less tables means less tips for me. And then, to be fair, I always feel like I need to give at least some of the tips I do get to the new trainee because they do assist a little bit. The bottom line is, every time you want me to be the trainer, I end up paying for it! That's not fair!"

Assume you were Demonte. Do you agree that the training plan for the new server is unfair to June? What could you do to ensure that the new employee gets the best training possible, but in a way that makes June happy to serve as the new employee's trainer?

Key Terms

Employer of choice
Mandatory benefit
Voluntary benefit
Open-door communication (policy)
Internal search (employee)
Employee referral
External search (employee)

Application
Equal Employment Opportunity Commission (EEOC)
Negligent hiring
Negligent retention
Training
Train-the-trainer (program)

Standard operating procedure (SOP)
Training plan
Training lesson
Training handbook
Pretest/post-test evaluation
Objective test
Mystery shopper

Operator's 10-Point Tactics for Success Checklist

Evaluate your need for, and the current status of, each of the following operational tactics. For those tactics you think are important, but not yet in place, develop an action plan for its implementation including who will be responsible for the tactic's completion and the target date by which it should be completed.

Tactic	Don't Agree (Not Done)	Agree (Done)	Agree (Not Done)	Who Is Responsible?	Target Completion Date
				If Not Done	
1) Operator recognizes that compensation, benefits, scheduling, and working conditions are all of prime importance when recruiting new team members.	____	____	____		
2) Operator understands the importance to success of internal recruiting when it is appropriate.	____	____	____		
3) Operator understands the importance to success of traditional and internet-based employee recruiting efforts.	____	____	____		
4) Operator ensures all job applicants complete a legally sound application for employment.	____	____	____		
5) Operator recognizes that the questions to be asked in an employment interview must focus on an applicant's job skills and nothing else.	____	____	____		
6) Operator has determined the role of pre-employment testing and background checks/references in their employee selection process.	____	____	____		

(Continued)

Tactic	Don't Agree (Not Done)	Agree (Done)	Agree (Not Done)	If Not Done	
				Who Is Responsible?	Target Completion Date
7) Operator has identified the federal, state, and local documentation required for all new employees.	____	____	____		
8) Operator has reviewed the personal characteristics of effective trainers before selecting those who will conduct their training.	____	____	____		
9) Operator can identify the five key steps to be taken in the creation of effective training programs.	____	____	____		
10) Operator recognizes the importance of training evaluation before, during, and after the delivery of training.	____	____	____		

9

Managing Food and Beverage Production and Service

What You Will Learn

1) The Importance of Managing Food and Beverage Production
2) How to Calculate Cost of Sales
3) The Importance of Managing Food and Beverage Service
4) How to Manage Service Recovery

Operator's Brief

In this chapter, you will learn that properly managing the production of your food and beverage menu items is extremely important to the success of your foodservice operation. One key aspect of effective management during production relates to using standardized recipes that are required for every menu item sold. In many cases, however, you will need to adjust a standardized recipe to produce the number of servings you want to prepare. In this chapter, you will learn how to adjust (scale) standardized recipe yields using a recipe conversion factor (RCF).

In this chapter, you will also learn how to properly calculate your operation's actual cost of sales; the cost of the products used to generate your food and beverage revenue.

The effective control of service-related procedures and their costs is just as important as managing product costs. In this chapter, you will learn that, in many ways, the challenges of professional service delivery management are even greater than those of product management. The reason: Management of guest service can be challenging because a guest's view of quality service differs from that of a foodservice operator. In this chapter, you will learn about these differences and their impact on your service management efforts.

(Continued)

Foodservice operators addressing their service-related efforts most often find that these efforts vary in large part based on the method used to provide service to their guests. Therefore, we will learn about the identification and control of service-related costs for operators providing:

✓ On-site service
✓ Drive-thru and off-site service

Despite your best efforts, your operation will likely experience some service deficiencies. Then you must undertake appropriate service recovery efforts to regain guest satisfaction. These efforts must be addressed for those guests who are on-site and for those who used drive-thru service or pick-up service and are no longer on-site.

CHAPTER OUTLINE

Successful Production Management
    Managing Food Production
    Managing Beverage Production
Controlling Production Costs
    The Importance of Standardized Recipes
    Calculating Cost of Sales
Successful Service Management
    The Guest's View of Quality Service
    The Operator's View of Quality Service
Controlling Service Costs
    On-Site Service
    Drive-Thru and Off-Site Delivery Service
Managing Service Recovery
    On-Site Service Recovery
    Off-Site Service Recovery

Successful Production Management

When professional procedures are implemented to properly purchase, receive, and store products, foodservice operators will be in a good position to manage the production of the menu items they sell.

For those operators who sell both food and alcoholic beverage products, these efforts must focus on:

✓ Managing food production
✓ Managing beverage production

Managing Food Production

Effective management of the production process is essential for success. In most cases, managing the production process in a foodservice operation means addressing four key areas:

✓ Waste
✓ Overcooking
✓ Over-portioning
✓ Improper carryover utilization

Waste

Product losses from food (and beverage) waste are one example of excessive food costs. Some waste may be easy to observe such as when an employee does not use a rubber spatula to remove all salad dressing from a 1-gallon jar. However, acts such as the improper work of a salad preparation person can yield excessive amounts of trim waste and, as a result, create a higher portion cost (see Chapter 7) for each salad sold.

Portion cost is the amount it costs an operation to produce one serving of a menu item. It is one important factor in establishing the selling prices of menu items, but it is also important for the control of production costs. Therefore, the ability to calculate a portion cost correctly is an important management skill.

In some cases, the calculation of a portion cost is very easy. For example, assume an operator sells fresh apples. The operator buys apples in 10-pound bags that contain 30 apples. Each bag costs $18.90. The operator can easily calculate the portion cost for one apple:

$$\frac{\$18.90 \text{ apple cost per bag}}{30 \text{ apple portions per bag}} = \$0.63 \text{ portion cost}$$

In other situations, operators utilize the cost required to produce a standardized recipe (see Chapter 1) when they calculate their portion costs.

For example, assume an operator sells pecan pie. The operator's standardized recipe for pecan pie produces seven pies. Each pie will be cut into eight pieces. Thus, the standardized recipe produces 7 pies × 8 portions per pie = 56 pie portions.

If the cost of producing the standardized recipe, including the pastry for the crust, pecans, eggs, brown sugar, salt, vanilla, milk, and nutmeg is $50.40, the operator could again use the same portion cost formula to calculate the cost of producing one slice of pecan pie:

$$\frac{\$50.40 \text{ standardized recipe cost}}{56 \text{ pie portions produced}} = \$0.90 \text{ portion (slice) cost}$$

Most items served by restaurateurs should always be produced by following standardized recipes. When this is done, it is easy to see why adherence to proper recipe production and portioning is critical to accurately calculating and controlling an operation's portion costs.

Foodservice operators can best control food production costs by consistently demonstrating their concern for the value of food products. Each employee should realize that wasting products directly affects the operation's profitability and the employees' own economic well-being.

In most cases, food waste results from poor training or an operator's inattentiveness to detail. Unfortunately, some operators and employees feel that small amounts of food waste are unimportant. However, effective operators know that a primary goal in reducing production waste should be to maximize product utilization.

Overcooking

Cooking is the process of exposing food to heat. Excessive cooking, however, most often results in reduced product volume regardless of whether the item being cooked is street taco filling or vegetable soup. The reason: Many foods have high moisture contents, and heating usually results in moisture loss. To minimize this loss, cooking times and methods included in standardized recipes must be calculated and followed.

Excess heat is the enemy of both well-prepared foods and an operator's cost control efforts. Too much time on the steam table line or in the holding oven extracts moisture from products. It also reduces product quality and yields fewer portions that are available for sale. The result is unnecessarily increased portion costs and total food costs.

To control product loss from overcooking, operators must strictly enforce standardized recipe cooking times. This is especially true for high moisture content items such as meats, soups, stews, and baked goods. It is, therefore, advisable to provide kitchen production personnel with small, easily cleanable timers and thermometers for which they are responsible. These can help substantially in reducing product losses from overcooking. It is also important to note that improperly cooked foods typically result in lower product quality, and this results in decreased guest satisfaction.

Over-portioning

Perhaps no other area of food production management is more important than the control of portion size. There are two reasons for this. First, over-portioning increases operating costs and may lead to mismatches in production schedules and anticipated demand. For example, assume 100 guests are expected, and 100 products to be served to them are produced. However, over-portioning causes the operation to be "out" of the product after only 80 guests were served. The remaining

20 guests will not receive portions because over-portioning allowed "their" food to be served to other guests.

Secondly, over-portioning must be avoided because guests want to believe they receive fair value for what they spend. If portions are large one day and small the next, guests may believe they have been cheated on the second day. Consistency is a key to operational success in foodservice, and guests want to know exactly what they will receive for the money they spend.

In many cases, inexpensive kitchen tools are available to help employees serve the proper portion size. Whether these tools consist of scales, food scoops, ladles, dishes, or spoons, employees require an adequate number of easily accessible portion control devices and use them consistently. The constant checking of portion sizes served is also an essential management task. When incorrect portion sizes are noticed, they must be promptly corrected to avoid product cost increases.

Improper Carryover Utilization

Most foodservice operators want to offer the same broad menu to the day's last guest as was offered to its first guest. It is inevitable that some prepared food will remain unsold at the end of the shift (this food is referred to as "carryovers" or "leftovers").

In some segments of the foodservice industry, carryovers are a potential problem area, but less so in other operations. Consider the operator of an upscale ice cream shop. At the end of the day, any unsold ice cream is held in a freezer until the next day with no measurable loss of either product quantity or quality.

Contrast that situation, however, with a full-service cafeteria. If closing time is 8:00 p.m., management wishes to have a full product line, or at least some of each menu item, available to the guest who walks in the door at 7:45 p.m. Obviously, in 15 more minutes, many displayed items will become carryovers, and this cannot be avoided. An operator's ability to effectively integrate carryover items on subsequent days can make the difference between profits and losses in some operations.

Carryover items that can be reused should be properly stored and labeled so the items can be found and reused easily. Most operators find that requiring foods to be properly stored and labeled in clear plastic containers help to manage these procedures.

Managing Beverage Production

In its simplest, but also its least desired form, beverage production can consist simply of a bartender who **free-pours** drinks.

Free-pouring occurs when a bartender makes a drink by pouring liquor from a bottle without

Key Term

Free-pour: Pouring liquor from a bottle without measuring the poured amount.

carefully measuring the poured amount. In a situation such as this, it is very difficult to control beverage production costs. At the other end of the control spectrum are automated total bar systems that are extremely sophisticated and costly control devices. Most foodservice operations will be operating under one of the following beverage product control systems:

✓ Free-pour
✓ Jigger pour
✓ Metered pour
✓ Beverage gun

The specific control system used will be based on the amount of control operators feel are appropriate for their own beverage businesses.

Free-Pour

The lack of control resulting from free-pouring alcohol is significant. It should never be allowed when preparing most drinks bartenders serve. It is appropriate in some settings, however, for example, in wine-by-the-glass sales. In this situation, the wine glass itself serves as a type of product control device. Large operations, however, may even elect to utilize an automated dispensing system for their "wines by the glass."

Jigger Pour

A **jigger** is a device used to measure alcoholic beverages that is typically marked in ounces and fraction of ounce quantities.

Jiggers are inexpensive to buy and use, so this control approach is inexpensive. It is a good system to use in remote serving locations such as a pool area, beach, guest suite, or banquet room. A disadvantage is that there is still room for employee overpouring and the potential for bartender fraud.

Key Term

Jigger: A small cup-like bar device used to measure predetermined quantities of alcoholic beverages. These items are usually marked in ounce and portions of an ounce. Examples: 1 ounce or 1.5 ounces.

Metered Pour

Some operators control their beverage production costs using a pour spout designed to dispense a predetermined (metered) amount of liquor each time the bottle is inverted.

Pour spouts are inserted into bottles and are available to dispense a variety of different quantities of beverage. When using a metered pour spout, the predetermined portion of product is dispensed whenever the bartender serves the product.

Beverage Gun

In some large operations, beverage guns are connected directly to liquor bottles or other liquor containers of various sizes. The gun may be activated by pushing a mechanical or electronic button built into the gun. In this situation, a bartender will, for example, push a gin and tonic button on a gun device. This, in turn, will result in dispensing a predetermined amount of both gin and tonic. Although the control features built into gun systems are many, their cost, lack of portability, and possible negative guest reactions can be limiting factors in their selection.

Controlling Production Costs

The effective control of production costs is among the most important tasks of foodservice operators. Two essential activities undertaken by successful operators to help control their production costs are the use of standardized recipes and the accurate calculation of cost of sales.

The Importance of Standardized Recipes

Standardized recipes (see Chapter 1) are essential to any serious effort to produce consistent, high-quality food and beverage products at a known product cost. All items on a foodservice operation's menu should be based on and follow a standardized recipe. When foodservice operators use standardized recipes, their staff will produce a product close to identical in yield and flavor whenever made regardless of who follows the recipe.

Adjusting standardized recipes using a process known as **scaling** means changing the number of servings a recipe produces by multiplying (to increase) or dividing (to decrease) the recipe ingredient quantities to match actual production needs.

Key Term

Scaling (recipe): The process of adjusting the yield of a standardized recipe.

To create an initial standardized recipe, it is always best to begin with a recipe of proven quality. For example, an operator may have a recipe designed to produce 10 portions but wants to expand it to yield 100 portions. In these situations, it may not be possible to simply multiply the amount of each recipe ingredient by 10.

A great deal has been written regarding various techniques used to expand recipes. Computer software designed for that purpose is also readily available to foodservice operators. Generally, any menu item that can be produced in quantity can be standardized in a recipe form.

When adjusting standardized recipes to produce a greater or fewer number of portions, it is important that recipe modifications be properly made. When adjusting recipes for proper quantity (number of servings desired to be produced), foodservice operators utilize a **recipe conversion factor (RCF)**.

When using the RCF method to adjust a recipe's yield, operators use the following formula to arrive at the appropriate RCF.

Key Term

Recipe conversion factor (RCF): A mathematical formula that yields a number (factor) operators can use to convert a standardized recipe with a known yield to the same recipe that produces a desired yield.

$$\frac{\text{Desired Recipe Yield}}{\text{Current Recipe Yield}} = \text{Recipe Conversion Factor (RCF)}$$

To illustrate, assume a standardized recipe currently yields 50 portions, but the number of portions an operator desires is 125. In this case, the RCF formula will be:

$$\frac{125 \text{ desired portions}}{50 \text{ current portions}} = 2.5 \text{ Recipe Conversion Factor (RCF)}$$

In this example, 2.5 is the RCF. To produce 125 portions, the operator multiplies the amount of each ingredient in the original (current) standardized recipe by 2.5 (the RCF) to arrive at the required amount of that ingredient. One good way to remember the principles of recipe yield conversion is to recognize that, if a recipe's yield is being *increased*, the recipe conversion factor (RCF) will always be *greater than 1.0*. If a recipe's yield is to be decreased, the RCF will always be *less than 1.0*.

Find Out More

Professional foodservice operators calculate recipe conversion factors (RCFs) for use in increasing and decreasing the size of their standardized recipes. Professional bakers accomplish the same task when they use specialized "baker's math" (sometimes referred to as baker's percentage) ratios to adjust the recipe size (or formulas).

When using baker's math, bakers apply the following formula:

$$\frac{\text{Ingredient Weight}}{\text{Flour Weight}} \times 100 = \text{Ingredient Percentage}$$

Essentially, when using baker's math, a baker calculates the total amount of flour needed in a baking formula. Each formula ingredient's weight is then divided by the weight of the flour to determine that ingredient's percentage of the flour's weight.

For example, if a bread formula makes 8 loaves and calls for 5 pounds of flour and 1 pound of sugar, the sugar's ingredient percentage would be calculated as

$$\frac{1\,\text{pound sugar}}{5\,\text{pounds flour}} \times 100 = 20\% \text{ sugar}$$

When the formula's sugar's percentage is known, it becomes easy for the baker to adjust the formula's yield for the desired number of loaves.

To learn more about methods bakers use to adjust their formulas (standardize recipes), enter "using baker's math" in your favorite search engine and view the results.

Technology at Work

Several good companies offer software programs that ease the process of standardized recipe development and necessary measurement conversion. The reason: Many recipes are produced using metric measurements rather than measurements in the imperial measurement system. Such programs help operators manage their standard recipe files, and they can also:

Calculate total recipe costs
Calculate per portion recipe costs
Help create new recipes
Plan menus
Conduct nutritional analysis of recipes
Track critical allergen information
Monitor product inventory levels in real time

To examine some of the innovative and increasingly essential standardized recipe-related cost control tools, enter "standardized recipe development software" in your favorite search engine and review the results.

Calculating Cost of Sales

All foodservice operators must be able to calculate their actual cost of sales. It is important to recognize that the cost of sales as shown on an operation's income statement (see Chapter 1) for an accounting period is most often *not* equal to the amount of food and beverage purchases in that same accounting period. In fact, foodservice operators must use a very specific process to accurately calculate their cost of sales and cost of sales percentages for both food and beverages.

Foodservice operators must know their actual cost of food sold during an accounting period. The cost of food sold formula foodservice operators use to calculate their costs is shown in Figure 9.1.

While it is commonly referred to as "cost of food sold," or "cost of sales," an operation's cost of food sold is actually the dollar amount of all food sold *plus* the costs of any food that was thrown away, spoiled, wasted, or stolen. To best understand and use the cost of food sold formula properly, operators must fully understand each of its individual parts.

Key Term

Beginning inventory: The monetary value of all products on hand at the start of an accounting period. Beginning inventory is determined by completing a physical inventory.

Beginning Inventory

When calculating cost of food sold, **beginning inventory** is the monetary value of all food on hand at the beginning of an accounting period (see Chapter 2).

An operation's beginning inventory can only be accurately determined by completing a **physical inventory**: an actual count and valuation of all foods in storage and in production areas. Some operators take a physical inventory monthly, and others weekly or even daily to determine their inventory amounts.

If, when taking a physical inventory, products are missed or undercounted, an operation's food

Key Term

Physical inventory: An inventory control system tool in which an actual (physical) count and valuation of all product inventory on hand is taken at the close of an accounting period.

Beginning Inventory

PLUS

Purchases

= Food Available for Sale
LESS

Ending Inventory

= Cost of Food Consumed
LESS

Employee Meals

= Cost of Food Sold

Figure 9.1 Formula for Cost of Food Sold

costs will ultimately appear higher than they actually are. If, on the other hand, an operator erroneously overstates the value of products in inventory, their product costs will appear artificially low.

It is important that operators take accurate physical inventories. To calculate food (or beverage) costs, they should recognize that the beginning inventory for an accounting period is always the **ending inventory** amount from the previous accounting period. For example, an operation's ending inventory on December 31 will become the operation's beginning inventory for January 1 of the next year.

Key Term

Ending inventory: The monetary value of all products on hand at the end of an accounting period.

Purchases

Purchases are the cost of all food purchased during the accounting period. The purchases amount is determined by properly adding the amounts on all food delivery invoices and any other bills for food products purchased during the accounting period being analyzed.

Food Available for Sale

Food available for sale is the sum of the beginning inventory and the purchases made during a specific accounting period. Some operators refer to food available for sale as "goods available for sale."

Ending Inventory

Ending inventory refers to the dollar value of all food on hand at the end of the accounting period. It also must be determined by completing an accurate physical inventory.

Cost of Food Consumed

The cost of food consumed is the actual dollar value of all food used (consumed) by the foodservice operation. Again, note that this amount is not just the value of all food sold. Instead, the number includes the value of all food no longer in the establishment due to sale, spoilage, waste, or theft. Cost of food consumed also includes the cost of any complimentary meals served to guests and the value of any food (meals) eaten by employees.

Employee Meals

The money spent to provide employee meals (if any) is actually a labor-related, not food-related, cost. Free or reduced-cost employee meals are a benefit frequently offered by employers in the same manner as the operation may provide medical insurance or paid vacations. Therefore, the value of this benefit, if provided, should not be recorded as a cost of food. Instead, the dollar value of all food eaten by employees is *subtracted* from the cost of food consumed and then *added* to the cost of labor on the income statement to accurately reflect an operation's actual cost of food sold.

Cost of Food Sold

An operation's cost of food sold (or "cost of goods" sold) is the actual amount of all food expenses incurred by the operation *minus* the cost of employee meals. It is not possible to accurately determine this cost unless a beginning physical inventory has been taken at the start of an accounting period. It must then be followed by another physical inventory taken at the end of the same accounting period.

Calculating actual cost of food sold on a regular basis is important because it is not possible to monitor or improve product cost control efforts unless an operator first knows what the food costs are. In nearly all operations, cost of food sold is calculated on at least a monthly basis. The reason: it is reported on the operation's income statement (P&L). In many operations, these same costs may be calculated on a weekly or even daily basis.

The proper computation of an operation's beverage cost of sales is identical to that of a food cost percentage with one important difference. Typically, there is no equivalent for employee meals because the consumption of alcoholic beverage products by working employees should be strictly prohibited.

Therefore, "employee alcoholic drinks" would never be considered as a reduction from overall beverage cost. However, like the cost of food sold, an operation's cost of beverage sold is the dollar amount of all alcoholic beverage products sold as well as the costs of all beverages that were given away, wasted, or stolen.

Managing and controlling product costs in both food and beverage production is important to all foodservice operators. However, their product cost control efforts do not end there. They must also be extremely concerned about controlling costs during customer service. This is true regardless of whether an operator serves most of its products on-site, through a drive-thru window, with customer carry out, and/or by delivery.

What Would You Do? 9.1

"How many dozen should I put in the proofer?" asked Ezra, the new baker at the Old Country Cafeteria.

Christopher was the day shift operations manager and, unfortunately, he did not know how to answer Ezra's question. What Ezra wanted to know was simple enough: "How many dozen rolls should be placed in the proofer to anticipate the night's dinner business?"

The problem was that the frozen dinner roll dough used at the Old Country Cafeteria needed to proof (be allowed to rise) for at least two hours before baking for 15 minutes. If too many rolls were proofed, they would not be needed, but they would still have to be baked and made into bread dressing or even discarded.

If too few dozen rolls were proofed, and the night was busier than anticipated, they would run out of "Fresh Baked Rolls" (one of the operation's signature items). Christopher knew that the night manager would be really upset. It was a daily guess, and sometimes Christopher missed the guess!

Assume you were the owner of the Old Country Cafeteria. How important to your cost production efforts do you think it would be for Christopher to have accurate information about the night's sales forecast? Would you prefer that Christopher over-estimate or under-estimate the number of rolls he placed in the proofer? Explain your answer.

Successful Service Management

To best understand the importance of managing service in a foodservice operation, it is essential to recognize that a foodservice operator is both a manufacturer and a retailer. A professional foodservice operator is unique because all functions of a product's sale, from menu development to guest service, can be the responsibility of the same person.

Foodservice operators must secure raw materials (menu ingredients), produce a product, and sell it—often under the same roof. Few other managers require the breadth of skills required to be effective foodservice operators. Since operators are in the service sector of business, many aspects of management are more challenging for them than for their manufacturing or retail management counterparts.

The face-to-face guest contact in the foodservice industry requires operators to assume the responsibility of standing behind their own work and that of their staff. This often occurs in a one-on-one situation with the ultimate consumer (end user) of an operation's products and services.

The management task checklist in Figure 9.2 shows some areas in which foodservice operators, manufacturing managers, and retailing managers differ in their responsibilities.

In addition to their role as food factory supervisors, foodservice operators must also manage service because, if they fail to perform this vital role, their business may perform poorly or can even cease to exist. Managing service can become particularly challenging when it is recognized that foodservice guests most often view quality service very differently from foodservice operators.

Task	Foodservice Operator	Manufacturing Manager	Retail Manager
1. Secure raw materials	Yes	Yes	No
2. Manufacture product	Yes	Yes	No
3. Market to end user	Yes	No	Yes
4. Sell to end user	Yes	No	Yes
5. Reconcile problems with end users	Yes	No	Yes

Figure 9.2 Management Task Checklist

The Guest's View of Quality Service

Experienced foodservice operators know a guest's view of service quality is impacted each time the guest experiences a **moment of truth**: any point in time where guests draw their own impression about a foodservice operation's product or service quality.

Foodservice operators must recognize that moments of truth can occur even before a foodservice guest arrives at an operation. For example, if a foodservice operation's marketing efforts result in a potential guest visiting the operation's website to place an online order, then the website must be easy to use and functional. If it is not, guests will likely conclude that the operation is not service-oriented. In this digital marketing example, the guest's moment of truth is most impacted by the actions of management when it developed the operation's website and not the operation's actual ability to deliver high-quality menu items in a timely manner.

While each foodservice operation is unique, guests in many operations will assess service quality as an operation's staff undertakes key activities in the following areas:

1) Greeting guests
2) Taking guests' orders
3) Delivering guests' orders
4) Serving guests as they enjoy their meals
5) Collecting payment

Foodservice guests increasingly assess the quality of an operation's service not through the moments of truth they personally experience, but rather through the

Key Term

Moment of truth: A point of interaction between a guest and a foodservice operation that enables the guest to form an impression about the business.

experience of others. Potential guests read about, and, in some cases, can even view the experiences of others as they search user-generated content (UGC) sites (see Chapter 4). The reason: their efforts frequently provide useful information about an operation's service (and product) quality as described by an operation's previous guests.

Operators frequently recognize the importance of ratings and user reviews to their digital reputations. They should continually monitor popular UGC review sites to see how their own guests have reviewed and rated their facilities for service (and product) quality. Note: Detailed information about how best to respond to negative UGC reviews is presented later in this chapter.

The Operator's View of Quality Service

All foodservice operators want to provide good service to their guests, and most operators believe they do so. These operators tend to view quality service not in terms of moments of truth, or even guest reviews, but rather they are concerned about three key service-related factors:

✓ Speed
✓ Accuracy
✓ Professionalism

Speed

In many foodservice operations, management believes their guests most often equate quality of service with speed of service. These managers recognize that guests are not likely, for example, to complain that their drive-thru orders were prepared too quickly! Guests will, however, complain if drive-thru lanes are long and/or if drive-thru order delivery times are excessive.

Similarly, in a table service restaurant, guests will not likely complain about being seated or served too quickly. There may be complaints, however, if guest wait times are excessively long. This is especially true if guests have made a reservation in advance and are not seated close to their reservation time.

Knowledgeable operators know that monitoring drive-thru times for operations providing this service, and **ticket times** for operations providing on-site dining are good ways to monitor service speed. Short service times are preferable to long service times, and desired standards for these times should be established and continually monitored.

Key Term

Ticket time: The amount of time required to fill a guest's order.

The actual ticket time standards appropriate for an individual operation vary based on the menu items it sells and the service levels it offers. However, some generally acceptable time-related standards common in the foodservice industry include:

✓ On-site dining:
  - Guest greeting and seating: 2–5 minutes from arrival
  - Drink service: 1–3 minutes from seating; 4–6 minutes from ordering
  - Appetizer service: 5–10 minutes from ordering
  - Entrée service:
    o Lunch: 8–12 minutes from ordering
    o Dinner: 12–15 minutes from ordering
  - Dessert service: 3–5 minutes from ordering
✓ Drive-thru service: 4–6 minutes from order placement to order delivery
✓ Carry out (pick-up) service: 15–30 minutes from order placement to order pickup
✓ Third-party delivery: 30–60 minutes from order placement to order delivery

---

Find Out More

Foodservice guests do not like to wait an excessively long time to be served. This is especially true for guests who are utilizing a foodservice operation's drive-thru service.

Interestingly, drive-thru times in foodservice operations have been increasing in recent years rather than declining. Part of this challenge occurs because quick-service restaurants continually expand their menus as they seek to attract more guests. While this can be positive, it can also increase drive-thru times as guests may take longer to select their items from an expanded list of alternatives.

Average drive-thru times for foodservice operations are regularly measured and publicized. The variation in drive-thru ticket times can be significant. For example, a recent survey found that drive-thru times at Taco Bell averaged 221.99 seconds (the shortest ticket time of the operations surveyed), while Chick-fil-A averaged 325.47 seconds (the longest ticket time of the operations surveyed).[*]

The best operators monitor changes in average ticket times in the foodservice industry to gain insight on current strategies to reduce these times. For more information, enter "tips for reducing drive-thru ticket times in restaurants" in your favorite search engine and view the results.

[*] https://www.delish.com/food-news/a41648070/restaurants-drive-thru-wait-times/ retrieved February 28, 2023.

Accuracy

From an operator's perspective, accuracy is second only to speed in helping to ensure guest satisfaction. Certainly, when guests do not receive the menu items they have ordered and/or when the prices charged for these items are incorrect, guest satisfaction levels will be reduced.

In addition, when the wrong menu items are produced and presented to guests and are not accepted because they are not what was ordered, product-related costs rise due to product waste. This is because menu items that are incorrectly prepared and delivered to a guest cannot typically be served to other guests.

To help minimize guest dissatisfaction because of order inaccuracy, servers must be carefully trained in order taking. In many cases, it is a good idea to repeat a guest's order fully before submitting the order to the operation's production staff for item preparation.

Professionalism

Experienced foodservice operators know that exhibiting professionalism in service delivery constitutes a variety of factors. These include providing facilities that are clean, properly lit, and maintained at a temperature that permits comfortable dining.

Staff-related factors that impact guests' views of an operation's professionalism include wearing clean uniforms, practicing appropriate personal hygiene, and projecting an image that reflects well on the operation.

When servers are polite, friendly, and demonstrate genuine concern for guest satisfaction, the operation projects the professionalism that guests expect and that will lead them to believe service levels are good. If order takers or servers are rude, appear rushed, or seem unconcerned about guest satisfaction, guests may respond by feeling that service levels are poor, even when speed and accurate service are acceptable.

Controlling Service Costs

In many cases, foodservice operators addressing their service-related issues will focus the greatest amount of attention on labor-related costs. It is true that front-of-house labor costs for positions such as hosts, servers, order takers, bussers, and the like constitute a significant portion of an operation's service costs. Other service-related costs, however, must be controlled as well.

It is important to recognize that the USAR format for an income statement (see Chapter 2) does not include a specific line for "service-related" costs. Rather, these costs are listed in a variety of designated cost categories. For example, the cost of live music provided to entertain guests while dining is listed in "Music and

Entertainment." Likewise, the cost of providing controlled temperature conducive to guest dining will be listed under "Utilities."

As a result, one good way to review specific areas in which service-related cost control efforts must be directed is to consider how an operation delivers its service. Broadly speaking, these can be viewed in two ways:

1) On-Site Service
2) Drive-thru and Off-Site Delivery Service

On-Site Service

The control of service when providing indoor dining to guests includes managing expenses related to the maintenance of an appropriate **ambience**. The environment created by an operation's heating, ventilating, and air conditioning (HVAC) system and the lighting provided are also an important aspect of these costs.

Key Term

Ambience: The character and atmosphere of a foodservice operation.

Additional on-site dining costs that must be controlled include the costs of linen and napkins (when those items are in use), as well as flatware, glassware, and china. In many on-site dining restaurants, the cost of employee uniforms is also significant, and these expenses must also be properly managed and controlled.

Outdoor dining on-site, also known as "alfresco dining" or "dining alfresco," is increasing in importance in some areas. Unique service-related costs incurred when an operation provides outdoor dining for guests include the cost of operating appropriate HVAC units and providing appropriate lighting. Entertainment-related costs such as live music are also common when providing outdoor dining services.

As part of the shift to increased outdoor dining, some foodservice operators now provide outdoor **dining bubbles**. Also known as bubble tents or dining pods, these enclosed and controlled environments allow guests to be served in colder (or warmer) weather and can be placed on sidewalks or in parking areas.

Key Term

Dining bubble: A controlled environment used to house foodservice guests while they are dining outdoors.

While every on-site dining facility is unique, it is important to recognize that operators providing such services must comply with the Americans with Disabilities Act. Doing so often requires ramps and spaces to allow mobility aids to move freely.

Drive-Thru and Off-Site Delivery Service

Increasing numbers of foodservice operators offer their guests a drive-thru service option. While historically drive-thru service was provided primarily by quick-service restaurant (QSR) operators, today more operations provide this popular service delivery style.

While it is possible (but extremely difficult) to provide drive-thru service without a digital menu board (see Chapter 6), operators providing drive-thru delivery service incur the costs of creating and maintaining this needed equipment.

In a guest pickup service, guests receive their menu items from drive-thru windows or receive their ordered items from designated on-site carry out areas. In addition to providing appropriate staffing and space for pick up, foodservice operators offering this service are most concerned with proper menu item packaging and packaging costs.

The packaging of take away items is just as important to ensuring guest satisfaction with the service as is the quality of the menu items being packaged. To illustrate, assume a foodservice operation received an order for a three chicken taco meal. The tacos *could* be fully assembled and placed in a Styrofoam container and, if so, the food would technically be ready for pick up.

It is much better, however, if the chicken fillings were packaged in a hot container. Chilled items like sour cream, cheese, and shredded lettuce can be packaged in a cold container, and the warm tortillas are wrapped in aluminum foil. In the second packaging scenario, packaging costs are increased, however, even when the guest's pickup time is delayed product quality is optimized.

It is important that foodservice operators recognize that, even when menu items are suitable as a take-out item and are packaged properly, product quality can still decline. This is especially so, if these items are not held correctly until they are picked up by guests and/or by delivery drivers.

When utilizing operator direct delivery for off-site guests, menu items are delivered to guests by an operation's own delivery employees. Direct service-related costs when utilizing this approach include those of driver wages and benefits, vehicle costs, and fuel costs.

For guest orders that will be delivered in coolers or insulated bags, the costs of proper cleanliness, sanitization, and disinfection between uses will be incurred. Increasingly, some operators add tamper-evident labels (those that, when ripped or torn, alert guests that their order was tampered with), or tamper-proof boxes that can also be used for their own deliveries.

Technology at Work

As take-out meals and third-party delivery companies continue to grow in popularity, the need for appropriate take-out packaging continues to evolve.

Peace of mind for customers receiving delivered meals is important and, as a result, there has been increased demand for tamper-proof (or tamper-evident) packaging of delivered meals. This type of packaging helps ensure that food has not been "sampled" or otherwise adulterated by delivery drivers.

(Continued)

Tamper-proof packaging provides peace of mind to guests and foodservice operations. Guests can be assured that their delivered menu items are safe and unadulterated. Foodservice operators can utilize third-party delivery services with reduced fear of tarnishing their operation's name by unscrupulous delivery drivers. Tamper-proof containers typically cost more than those container types that do not offer such features, so this is a primary cost concern as well.

Innovative packaging for takeout and delivery foods continues to evolve. To learn more about the tamper-proof products resulting from this evolution, enter "tamperproof packaging for restaurant meal delivery" in your favorite search engine and review the results.

Third-party delivery costs are incurred when menu items are delivered to guests by one or more third-party delivery partners selected by a foodservice operation. The COVID pandemic that began in 2019 significantly increased guests' use of third-party delivery services for off-site guests. During the height of the pandemic, many foodservice operations were closed for indoor on-site dining and, also, many guests decided it was safer to eat in their homes than to dine out.

The basic business model of third-party delivery companies is relatively straightforward. Foodservice operators agree to have their operations listed on the app developed and marketed by the third-party delivery company, and the operator then provides a menu that includes item prices. Guests utilizing the apps place their orders with the third-party delivery company (not directly with the operation). Guest orders are then forwarded to the foodservice operation.

The third-party delivery company picks up and delivers guest orders and charges a commission to the foodservice operation for their services. Typical commission fees range from 20% to 30% of the dollar amount of the order. The remaining amount of the food order's value is then electronically transferred to the foodservice operation's bank account.

Those foodservice operators who partner with third-party delivery services typically do so with the belief their profitability will increase. However, despite increasing popularity, some foodservice operators seeking to control service costs and ensure profitability are concerned about the high costs of third-party delivery partnerships. For foodservice operators, the delivery fees paid to third-party delivery app operators reduce the amount of money to pay for food, labor, and other operating expenses required to produce the menu items that are delivered. As a result, the decision to provide off-site delivery services to guests via third-party delivery apps must be considered very carefully.

Managing Service Recovery

Even the best managed foodservice operations experience shortfalls in customer service and must plan for and use appropriate **service recovery** efforts.

The consideration of effective service recovery procedures is required for several reasons:

1) *Maintenance of good reputation:* By implementing effective service recovery strategies, foodservice operators take steps to ensure their guests are satisfied even when problems occur.
2) *Improved guest retention:* Guests who are satisfied with an operation's service levels are likely to return, even if they have experienced a service-related problem. Note: This is only true if the problem is addressed in a manner that is satisfactory to the guest.
3) *Reduced service recovery costs:* If front-of-house service staff know in advance what to do under a variety of different service recovery strategies, appropriate actions can be determined before they occur. Note: This helps ensure that costs associated with service recovery are known in advance and are appropriately incurred.

Specific service recovery steps in a foodservice operation can range from a sincere apology by a server or manager to monetary compensation to guests including, for example, the "**comping**" of all, or part, of a guest's bill.

Recovery steps related to food quality issues typically involve the replacement (comping), of a menu item that was, for example, overcooked, undercooked, or served at an incorrect temperature.

Service recovery steps related to service quality issues can best be viewed where they take place. In most cases, foodservice operators must consider the service recovery steps and resulting costs that are undertaken when guests are:

✓ On-Site
✓ Off-Site

On-Site Service Recovery

On-site service recovery means taking care of a service-related shortcoming while the affected guest is still on premises. Whenever possible, it is best to attempt to resolve a guest's problem immediately upon becoming aware of it. In some cases, a guest will raise an issue with a service staff member, and, in other cases, the guests may request to speak with the manager. In either situation, a prompt response to the guest's concern is essential.

The best method of dealing with a service-related failure is based on both the specific operation where it occurs and the cause of the service failure. In general, foodservice operators should train staff to select and use one or more of the following service recovery strategies (these are listed in terms of incremental cost).

1) **Sincere apology:** In some cases, a guest's complaint can be corrected merely by a sincere apology. For example, consider when a guest requests iced tea to be served with lemon. However, the beverage when served does not include a lemon slice. This guest complaint can normally be addressed simply by apologizing for the service error and assuring the guest that the proper item will be promptly delivered. Apologies cost an operator nothing and, in many cases, serve to properly demonstrate concern for a guest's service-related issue.

2) **Complimentary menu item:** In some cases, an operation can best address a guest's complaint by giving servers the power to "comp" a specific menu item. For example, a server may be authorized to offer a complimentary appetizer for a table of guests with a confirmed reservation who waited a long time past their reservation time to be seated.

3) **Guest check reduction:** If a menu item has been served, and it did not meet the quality standards the guest anticipated, service recovery can involve the elimination of the menu item from the guest's bill. This can be done whether the guest has consumed the menu item or not and is often a cost-effective way to resolve a service or product deficiency. Each operation must determine in advance which staff members will be allowed to make a guest check reduction and the upper dollar amount limits of such a reduction prior to seeking management approval.

4) **Guest check comp:** In the case of a particularly egregious service failure, an operator may resolve the issue by comping a guest's entire check. This service recovery strategy should only be undertaken with the consent of management.

5) **Credit for a future visit:** In some cases, it may be the best course of action to offer a guest a credit to be utilized on a *future* visit. When this approach is taken, a dollar amount should be established. For example, a guest might be given a gift certificate or gift card for $25 off their next meal. This service recovery strategy should only be undertaken with the consent of management.

Service recovery efforts implemented by a foodservice operator should recognize that the goal is to ensure that a guest who has experienced a negative moment of truth leaves the operation satisfied. When they do, they are more likely to return and are less likely to share their negative service experience with others or online.

Off-Site Service Recovery

Off-site service recovery efforts take place after a guest has left the foodservice operation, has picked-up their order, or received a delivery from the on-site operation or a **ghost kitchen**. In many cases, these efforts are undertaken in response to a negative review of an operation posted online. This demonstrates that problems solved immediately (on-site) lead to higher online evaluation scores because guests with corrected negative experiences are less likely to write about their problem when promptly resolved.

Negative reviews typically result in a reduction to the average "score" achieved by a foodservice operation. The average score achieved on a UGC review site (see Chapter 4) is important to guests deciding whether to visit a specific foodservice operation. Therefore, scores achieved by an operation directly affect the amount of business they will do. As a result, operators should regularly (several times weekly, if possible) respond to negative (and positive!) online reviews (see Chapter 5).

Key Term

Ghost kitchen: A foodservice operation that provides no on-site dine-in services and that prepares all of its menu items only for pick up or delivery to guests. Also known as a delivery-only restaurant, shadow kitchen, cloud kitchen, or virtual kitchen.

The ability to manage food and beverage production and service is critical to the success of a foodservice operation. In addition, the management of other operating costs, including especially the cost of labor, is also essential to success. The successful management of staffing costs and other operating expenses are so important to an operation's long-term success, these will be the sole topics addressed in the next chapter.

What Would You Do? 9.2

"We've got to fix this," said Clara, the dining room manager at Butterflies Bistro.
"Fix what?" asked Elijah, the Bistro's kitchen manager.
"This average score we're getting on the "DineInTown" website," replied Clara.
The "DineInTown" website featured guest reviews of local restaurants, and it was a popular site with many postings. The site allowed reviewers to rank restaurants on a scale of 1–5. (1 being a low score, and 5 being the highest score.)

(Continued)

"We are averaging a score of 3.8," said Clara. "I don't expect us to get all 5s, but we should be at least in the 4 plus range. The interesting thing is that we tend to get almost all 5s because reviewers say they love our food, and reviewers who give us a score of 1 almost always refer to slow service. These are mostly guests who visited us on a Friday or Saturday night. With lots of 5s and then a few 1s mixed in, we end up with a 3.8 average. That's what we need to fix."

Assume you were Clara. What do you think your guest reviews are telling this operation about its staffing patterns on Friday and Saturday nights? How important would it be for you to carefully manage reservation capacity and service and production staffing levels on these two busy nights?

Key Terms

Free-pour	Beginning inventory	Ambience
Jigger	Physical inventory	Dining bubble
Scaling (recipe)	Ending inventory	Service recovery
Recipe conversion factor (RCF)	Moment of truth	Comp
	Ticket time	Ghost kitchen

Operator's 10-Point Tactics for Success Checklist

Evaluate your need for, and the current status of, each of the following operational tactics. For those tactics you think are important, but not yet in place, develop an action plan for its implementation including who will be responsible for the tactic's completion and the target date by which it should be completed.

				If Not Done	
Tactic	Don't Agree (Not Done)	Agree (Done)	Agree (Not Done)	Who Is Responsible?	Target Completion Date
1) Operator recognizes the importance of controlling product costs during food and beverage production.	___	___	___		
2) Operator can calculate the recipe conversion factor (RCF) required to properly increase or decrease the yield of a standardized recipe.	___	___	___		

Tactic	Don't Agree (Not Done)	Agree (Done)	Agree (Not Done)	If Not Done Who Is Responsible?	Target Completion Date
3) Operator can accurately calculate their cost of food sold.	___	___	___		
4) Operator can accurately calculate their cost of beverage sold.	___	___	___		
5) Operator has carefully considered the importance of managing moments of truth when training staff to provide excellence in guest service.	___	___	___		
6) Operator recognizes the need to consider speed, accuracy, and professionalism when designing high-quality guest service programs.	___	___	___		
7) Operator has considered the specific service-related cost control challenges of providing on-site indoor and outdoor dining.	___	___	___		
8) Operator has considered and evaluated the specific service-related cost control challenges of providing off-site drive-thru delivery, guest pick up, and third-party delivery of menu items.	___	___	___		
9) Operator has communicated to service staff the progressive approaches to be taken to address service shortcomings when affected guests are still on-site.	___	___	___		
10) Operator recognizes the importance of monitoring UGC review sites to identify appropriate off-site service recovery strategies.	___	___	___		

10

Managing the Cost of Labor and Other Foodservice Expense

What You Will Learn

1) The Importance of Controlling Labor Costs
2) How to Evaluate Labor Productivity
3) The Importance of Controlling Other Operating Expenses
4) How to Compute Other Expense Costs: Percentage of Total Sales and Cost per Guest

---

Operator's Brief

In this chapter, you will learn about controlling labor costs, an expense that, for most foodservice operations, is second in importance only to that of product costs. In some operations, labor costs actually exceed the amount spent on food and beverage products.

The expenses of management, staff, and benefits comprise an operation's total labor costs. In addition to the cash amounts paid for these items, however, there are numerous noncash factors that can affect an operation's total cost of labor.

Examples include employee selection, training, the menu, and desired levels of service. Each of these noncash factors is examined in this chapter.

Productivity standards are needed to determine how much labor expense is needed to operate your business. A variety of productivity measures including sales per labor hour, labor dollars per guest served, and labor cost percentage are possible productivity measures. Each is addressed in detail in this chapter including an examination of how they are calculated and their strengths and weaknesses.

In addition to the effective control of products, service, and labor costs, foodservice operators must be concerned about other expenses required to operate their businesses. In this chapter, you will learn the importance of managing these other operating expenses.

In some cases, other expenses will be controllable by you, and in other cases these costs are considered noncontrollable. In this chapter, you will learn the differences between these two types of costs.

As you control other expense costs, you can employ two important cost control tools. These are the "other expense cost percentage" and the "other expense cost per guest." Both are important tools depending on the specific information you seek. In this chapter, you will learn how to calculate and assess these important other expense control metrics.

CHAPTER OUTLINE

The Importance of Labor Cost Controls

In most foodservice operations, labor is second only to product costs as the largest incurred expense and, in some operations, labor costs exceed food and beverage product costs. In years past, labor was relatively inexpensive. Today, however, the labor market is increasingly costly. To address these costs, foodservice operators must learn the scheduling and supervisory skills needed to optimize labor cost and to maximize staff effectiveness. They must also apply cost control skills to evaluate their efforts.

In some sectors of the foodservice industry, a reputation for long hours, poor pay, and undesirable working conditions have caused some high-quality employees to look elsewhere for more satisfactory careers. However, it does not have to be that way. When labor costs are adequately controlled, foodservice operators will typically have the funds necessary to create desirable working conditions and pay wages attractive to the best employees. In every business, including foodservice, better employees mean better guest service and, ultimately, better profits.

Total Labor Costs

Prime costs (see Chapter 2) in a foodservice operation are the sum of its product (food and beverage) costs and its total labor costs. Therefore, when total labor expense is properly controlled or reduced, prime costs are also reduced.

When utilizing the Uniform System of Accounts for Restaurants (USAR) (see Chapter 2) total labor cost included in an operation's prime cost calculations are the expenses for the operation's management, staff, and employee benefits, as shown in Figure 10.1.

Typically, any employee who receives a salary is considered management. A **salaried employee** receives the same income per week or month regardless of the number of hours worked.

Salaried employees are more accurately described as **exempt employees** because their duties, responsibilities, and levels of decisions make them "exempt" from the overtime provisions of the U.S. federal government's Fair Labor Standards Act (FLSA).

Exempt employees do not receive overtime pay for hours worked in excess of 40 per week, and they are expected by most foodservice operators to work the number of hours needed to adequately perform their jobs.

Key Term

Salaried employee: An employee who regularly receives a predetermined amount of compensation each pay period on a weekly or less frequent basis. The predetermined amount paid is not reduced because of variations in the quality or quantity of the employee's work.

Key Term

Exempt employee: Employees exempted from the provisions of the Fair Labor Standards Act. These workers typically are paid a salary above a certain level and work in an administrative or professional position. The U.S. Department of Labor (DOL) regularly publishes a duties test that can help employers determine who meets this exemption.

Figure 10.1 Components of Total Labor Costs (per USAR)

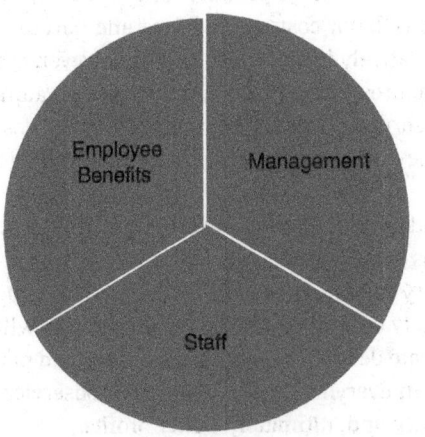

Staff costs in a foodservice operation generally refer to the gross pay received by employees in exchange for their work. For example, if an employee earns $18.00 per hour and works 40 hours per week, the gross paycheck (the employee's paycheck before any mandatory or voluntary payroll deductions) would be $720 ($18.00 per hour × 40 hours = $720). This gross amount is recorded as a staff cost.

Employee benefit costs are incurred in every foodservice operation. Mandatory benefits such as required contributions to Social Security are included as an employee benefit cost. Another example is voluntary contributions such as an employer's contribution to an individual employee's 401K retirement account.

The actual total amount of employment taxes and benefits paid by a specific operation can vary greatly. Expenses such as payroll taxes and contributions to workers' unemployment and workers' compensation programs are mandatory for all employers. Other benefit payments such as those for employee insurance and retirement programs are voluntary. They vary based on the benefits a foodservice operation offers its employees. As employment taxes and benefit costs increase, an operation's total labor cost increases, even its management and staff costs remain constant.

Technology at Work

Accounting for, and properly recording employment taxes due on, employee tipped income is an important task for many foodservice operators. Tip distribution software, also referred to as tip pooling or gratuity management software, automates the process of paying tips to tipped workers at the end of their shifts and reduces the need for physical cash payouts.

These tools eliminate hours of manual labor by tracking the number of hours worked and automatically distributing tips to workers' bank accounts. These tools can also reduce payroll burdens and instances of tip disparity due to errors or theft. They also expedite paying out tips to employees by connecting securely to their bank accounts or debit cards.

Additionally, most tip distribution software allows operators to manage tips across multiple locations and utilizes proper reporting features to reduce risks of noncompliance in tip payments.

To examine some of the offerings of companies that have developed software to assist in tip payment and proper recording, enter "tip distribution software in restaurants" in your favorite search engine and review the results.

Factors Affecting Total Labor Costs

For some foodservice operators, analysis of total labor costs only involves an assessment of the dollar amount in salaries and staff wages they pay, plus the cost of any benefits they provide.

Experienced foodservice operators, however, recognize that their total labor costs are directly affected by several other factors. Among the most important of these are:

1) Employee selection
2) Training
3) Menu
4) Service level provided

Employee Selection

Choosing the right employee for a vacant position in a foodservice operation is vitally important in developing a highly productive workforce. The best foodservice operators know that proper employee selection procedures go a long way toward establishing the kind of workforce that is both efficient and cost-effective. This involves matching the right employee with the right job, and the process begins with the development of the **job description**.

A written job description should be maintained for every position in every foodservice operation. From the job description, an appropriate **job specification** can be prepared.

Job descriptions and job specifications are important because they enable foodservice operators to hire only those employees who are qualified to do a job and do it well. The reason: Qualified workers can complete their tasks in a more cost-effective manner than will those who are not qualified.

Key Term

Job description: A statement that outlines the specifics of a particular job or position within a foodservice operation. It provides details about the responsibilities and conditions of the job.

Key Term

Job specification: A listing of the personal characteristics and skills needed to perform the tasks contained in a job description.

Training

Perhaps no area under an operator's direct control holds greater promise for increased employee productivity than effective training. In too many cases, however, training in the hospitality industry is poor or almost nonexistent. Highly productive employees are usually well-trained employees, and frequently employees with low productivity have been poorly trained. Every position in a foodservice operation should have a specific, well-developed, and ongoing training program.

Effective training improves job satisfaction and instills in employees a sense of well-being and accomplishment. It will also reduce confusion, product waste, poor service, and loss of guests. In addition, supervisors find that a well-trained workforce is easier to manage than one in which employees are poorly trained.

An operator's training programs need not be elaborate, but they must be consistent and continual. Foodservice employees can be trained in many areas. Skills training allows production employees to understand an operation's menu items and how they are best prepared. Service-related training may be undertaken, for example, to teach new employees how arriving guests should be greeted and properly seated, and/or how drive-thru orders should be entered in an operation's point-of-sale (POS) system.

It is important to recognize that training must be ongoing to be effective. Employees who are well-trained in an operation's policies and procedures should be constantly reminded and updated if their skill and knowledge levels are to remain high. Performance levels can also decline because of a change in the operational systems or changes in equipment used. When these changes occur, employees must be retrained. Effective training costs a small amount of time in the short-term, but these expenses can pay off extremely well in long-term savings.

Menu

A major factor in employee productivity is an operation's actual menu. The menu items to be served often have a significant effect on employees' ability to produce the items quickly and efficiently.

In most cases, the greater the number of menu items a kitchen must produce, the less efficient that kitchen will be. Of course, if management does not provide guests with enough choices, loss of sales may result. Clearly, neither too many nor too few menu choices should be offered. The question for operators most often is, "How many selections are too many?" The answer depends on the operation, its employees' skill levels, and the variety of menu items operators believe is necessary to properly attract and serve their guests.

It is extremely important that the menu items selected by management are those items that can be prepared efficiently and serviced well. When done, worker productivity rates will be high, and so will guest satisfaction.

Service Level Provided

The average quick-service restaurant (QSR) employee normally serves more guests in an hour than the fastest server at an exclusive fine dining restaurant. The reason: QSR guests desire speed, not extended levels of rendered services. In contrast, fine dining guests expect more elegant and personal service that is delivered at a much higher level, and this increases the number of necessary employees.

After operators fully recognize the noncash payment factors that impact their total labor costs, they must understand the different types of payment-related labor expenses to be incurred. Operators must also evaluate the productivity of their workers to determine if their labor dollars were well spent.

Technology at Work

In many foodservice operations, the cost of labor is equal to, or even exceeds, the operation's cost of sales. In a foodservice operation, the need for hourly paid employees can vary tremendously at different times of the day and on different days of the week. Therefore, developing and maintaining employee schedules is a critical management task.

The scheduling of hourly employees can be both time consuming and challenging as operators plan the number of employees who should work, and when they should work. This is especially so when the operation employs many part-time workers with varying scheduling needs.

Fortunately, some companies have designed software programs that assist operators in creating and quickly modifying hourly worker schedules. The best of these programs allows employees to access the schedule from their own smart devices. Then, there is no need to return to the operation to learn if staff have been added to or reduced from the work schedule.

Employee shift scheduling software saves time and ensures employees are always scheduled according to the operation's needs and the employees' personal preferences and availability. Basic employee scheduling programs are often offered at no cost, and those with more advanced features are available for lease.

To review the features of free-to-use employee scheduling tools, enter "best free scheduling software for restaurants" in your favorite search engine and review the results.

Assessment of Total Labor Costs

To determine how much labor expense is needed to properly operate their businesses, foodservice operators must determine how much work is to be done, and the amount of work each employee can accomplish. If too few employees are scheduled to work, poor service and reduced sales can result because guests may choose to go elsewhere in search of superior service levels. If too many employees are scheduled, staff wages and employee benefits costs will be too high, and the result will be reduced profits.

To properly determine the number of staff members needed, operators must have a good understanding of the **productivity** of each of their employees.

Key Term

Productivity (worker): The amount of work performed by an employee within a fixed amount of time.

There are several ways to assess labor productivity. In general, productivity is measured by calculating a productivity **ratio**.

The formula used to calculate a productivity ratio is:

$$\frac{\text{Output}}{\text{Input}} = \text{Productivity Ratio}$$

Key Term

Ratio: An expression of the relationship between two numbers. Ratios are computed by dividing one number by the other number.

To illustrate the use of this ratio, assume a foodservice operation employs 4 servers, and it serves 80 guests. Using the productivity ratio formula, the output is guests served, and the input is servers employed:

$$\frac{80 \text{ guests}}{4 \text{ servers}} = 20 \text{ guests per server}$$

This formula states that, for each server employed, 20 guests can be served. The productivity ratio is 20 guests to 1 server (20 to 1) or, stated another way, 1 server per 20 guests (1 to 20).

Productivity ratios can help an operator determine the answer to the key question, "How much should I spend on labor?" Foodservice operators must develop their own methods for managing payroll costs because every foodservice unit is different. Consider, for example, the differences between managing payroll costs incurred by a food truck operator and those required for a large banquet kitchen located in a 1,000-room convention hotel.

Although methods used to manage payroll costs may vary, payroll costs can and must be managed. While operators can use several ratios to assess their labor costs (worker productivity), in the foodservice industry the three most commonly utilized productivity ratios are:

1) Sales per labor hour
2) Labor dollars per guest served
3) Labor cost percentage

Sales per Labor Hour

It has been said that the most perishable commodity any foodservice operator can buy is the labor hour. When labor is not productively used, it disappears forever. It cannot be "carried over" to the next day as can an unsold head of lettuce or a slice of turkey breast. For this reason, some foodservice operators measure labor productivity in terms of the amount of sales generated for each labor hour used. This productivity measure is referred to as **sales per labor hour**.

Key Term

Sales per labor hour: The dollar value of sales generated for each labor hour used.

The formula used to calculate sales per labor hour is:

$$\frac{\text{Total Sales Generated}}{\text{Labor Hours Used}} = \text{Sales per Labor Hour}$$

When using this productivity measure, labor hours used is simply the sum of all labor hours paid for by an operation within a specific sales period. To illustrate, consider the operator whose four-week labor usage and the resulting sales per labor hour information is presented in Figure 10.2.

In this example, sales per labor hour ranged from a low of $19.50 in Week 1 to a high of $28.66 in Week 4. Sales per labor hour varies with changes in selling prices, but it will not vary based on changes in prices paid for labor. In other words, increases and decreases in the price paid per hour of labor will not affect this productivity measure. As a result, a foodservice operation paying its employees an average of $15.00 per hour could, using this measure for labor productivity, have the same sales per labor hour as a similar unit paying $20.00 for each hour of labor used. Obviously, the operator paying $15.00 per hour has paid far less for an equally productive workforce if the sales per labor hour used are identical in the two units.

Many operators like utilizing the sales per labor hour productivity measure because records on both the numerator (total sales) and the denominator (labor hours used) are readily available. However, depending on the recordkeeping system employed, it may be more difficult to determine total labor hours used than total labor dollars spent. This is especially true when large numbers of managers or supervisors are paid by salary rather than by the hour. Note: It is an operator's choice whether the efforts of both salaried workers and hourly paid staff should be considered when computing an operation's overall sales per labor hour. If the operator is consistent with this choice, sales per labor hour from different sales periods can be appropriately compared.

Week	Total Sales	Labor Hours Used	Sales per Labor Hour
1	$18,400	943.5	$19.50
2	21,500	1,006.3	21.37
3	19,100	907.3	21.05
4	24,800	865.3	28.66
Total	$83,800	3,722.4	$22.51

Figure 10.2 Four-Week Sales per Labor Hour

Labor Dollars per Guest Served

Some foodservice operators measure labor productivity in terms of the labor dollars spent for each guest served. This productivity measure is referred to as **labor dollars per guest served.**

The formula used to calculate labor dollars per guest served is:

$$\frac{\text{Total Cost of Labor}}{\text{Total Number of Guests Served}} = \text{Labor Dollars per Guest Served}$$

To illustrate the use of this ratio, consider the operator whose four-week labor cost and the resulting sales per labor hour information is presented in Figure 10.3.

In this example, the labor dollars per guest served for the four-week period would be computed as:

$$\frac{\$29,330}{\$4,190} = \$7.00$$

Note that, in this example, for three weeks (weeks 1–3) the operator provided guests with more than $7.00 of guest-related labor costs per guest served. However, in the fourth week, that amount fell to less than $6.00 per guest. This productivity measure, when averaged, can be useful for operators who find that their labor dollars expended per guest served are lower when their volume is high and higher when their volume is low.

The utility of labor dollars per guest served is limited because it varies based on the price paid for labor. Unlike sales per labor hour, however, it is not affected by changes in menu prices.

Week	Cost of Labor	Guests Served	Labor Dollars per Guest Served
1	$7,100	920	$7.72
2	8,050	1,075	7.49
3	7,258	955	7.60
4	6,922	1,240	5.58
Total	$29,330	4,190	$7.00

Figure 10.3 Four-Week Labor Dollars per Guest Served

Labor Cost Percentage

The most common measure of employee productivity in the foodservice industry is the **labor cost percentage**.

The formula used to calculate a labor cost percentage is:

Key Term

Labor cost percentage: A ratio of overall labor costs incurred relative to total revenue generated.

$$\frac{\text{Total Cost of Labor}}{\text{Total Revenue Generated}} \times 100 = \text{Labor Cost \%}$$

A labor cost percentage allows an operator to measure the relative cost of labor used to generate a known quantity of sales. It is important to realize, however, that different operators may choose slightly different methods of calculating this popular productivity measure.

Since a foodservice operation's total labor cost consists of management, staff, and employee benefit costs, some operators may calculate their labor cost percentage using only hourly staff wages, or staff wages and salary costs (but not benefit costs). This approach makes sense if an operator can directly control employee pay but not employee benefit costs. It is important to recognize, however, that when operators wish to directly compare their own labor cost percentage to that of other operations, both foodservice programs must have utilized the same formula.

Controlling the labor cost percentage is extremely important in the foodservice industry because it is often used to assess the management's effectiveness. If an operation's labor cost percentage increases beyond what is expected, management will likely be held accountable by the operation's ownership.

Labor cost percentage is a popular measure of productivity, in part, because it is so easy to compute and analyze. To illustrate, consider Roberta, a foodservice manager in charge of a casual service restaurant in a year-round theme park. The unit is popular and has a $20 per guest check average. Roberta uses only payroll (staff wages and management salaries) when determining her overall labor cost percentage. The reason: She does not have easy access to the actual amount of taxes and benefits provided to her employees. Roberta's own supervisor considers these labor-related expenses to be noncontrollable and, therefore, beyond Roberta's immediate influence.

Roberta has computed her labor cost percentage for each of the last four weeks using her modified labor cost percentage formula. Her supervisor has given Roberta a goal of 35% labor costs for the four-week period. Roberta feels that she has done well in meeting that goal. Figure 10.4 shows Roberta's four-week performance.

Week	Cost of Labor	Total Sales	Labor Cost %
1	$7,100	$18,400	38.6%
2	8,050	21,500	37.4
3	7,258	19,100	38.0
4	6,922	24,800	27.9
Total	**$29,330**	**$83,800**	**35.0**

Figure 10.4   Roberta's Four Week Labor Cost Percentage Report

Using her labor cost percentage formula and the data in Figure 10.4, Roberta's Labor Cost % is calculated as:

$$\frac{\text{Cost of Labor}}{\text{Total Sales}} \times 100 = \text{Labor Cost \%}$$

Or

$$\frac{\$29,330}{\$83,800} \times 100 = 35\%$$

Roberta did achieve a 35% labor cost for the four-week period. However, Monica, the supervisor, is concerned because several negative comments were received in week 4 about poor service levels in Roberta's unit. Some of these were even posted online, and Monica is concerned about the postings' potential impact on future visitors to the park's foodservice operations. When she analyzes the numbers in Figure 10.4, she noticed that Roberta exceeded her goal of a 35% labor cost in weeks 1 through 3 and then reduced her labor cost to 27.9% in week 4.

Although the monthly overall average of 35% is within budget, Monica knows all is not well in this unit. To achieve her assigned goal, Roberta reduced her payroll in week 4. However, the negative guest comments suggest that reduced guest service resulted from too few employees on staff to provide the necessary guest attention. As Monica recognized, one disadvantage of using an overall labor cost percentage is that it can hide daily or weekly highs and lows!

In Roberta's operation, labor costs were too high the first three weeks, and too low in the last week, but she still achieved her overall target of 35%. Roberta's labor cost of 35% indicates that, for each dollar of sales generated, 35 cents was paid to the employees who assisted in generating those sales. In some cases, a targeted labor cost percentage is viewed as a measure of employee productivity and, to some degree, management's skill in controlling labor costs.

Week	Original Cost of Labor	5% Pay Increase	Cost of Labor (with 5% Pay Increase)	Total Sales	Labor Cost %
1	$7,100	$355.00	$7,455.00	$18,400	40.5%
2	8,050	402.50	8,452.50	21,500	39.3
3	7,258	362.90	7,620.90	19,100	39.9
4	6,922	346.10	7,268.10	24,800	29.3
Total	29,330	1,466.50	30,796.50	83,800	36.8

Figure 10.5 Roberta's Four-Week Revised Labor Cost % Report (Includes 5% Pay Increase)

While it is popular, in addition to its tendency to mask productivity's highs and lows, the labor cost percentage has some limitations as a measure of productivity. Note, for example, what happens to this measure of productivity if all of Roberta's employees are given a 5% raise in pay. If this were the case, her labor cost percentages for last month would be calculated as shown in Figure 10.5.

Note that labor now accounts for 36.8% of each sales dollar, but one should realize that Roberta's workforce did not become less productive simply because they got a 5% increase in pay. Rather, the labor cost percentage changed due to a difference in the price paid for labor. When the price paid for labor increases, labor cost percentage increases. Similarly, when the price paid for labor decreases, the labor cost percentage decreases. Therefore, using the labor cost percentage alone to evaluate workforce productivity can sometimes be misleading.

Another example of the limitations of the labor cost percentage as a measure of labor productivity can be seen when selling prices are increased. Return to the data in Figure 10.5 and assume that Roberta's unit raised all menu prices by 5% effective at the beginning of the month. Figure 10.6 shows how this increase in her selling prices would affect her labor cost percentage.

Week	Cost of Labor	Original Sales	5% Selling Price Increase	Sales (with 5% Selling Price Increase)	Labor Cost %
1	$7,100	$18,400	$920	$19,320	36.7%
2	8,050	21,500	1,075	22,575	35.7
3	7,258	19,100	955	20,055	36.2
4	6,922	24,800	1,240	26,040	26.6
Total	29,330	83,800	4,190	87,990	33.3

Figure 10.6 Roberta's Four-Week Revised Labor Cost % Report (Includes 5% Increase in Selling Price)

Note that increases in selling prices (assuming no decline in guest count or changes in guests' buying behavior) will result in labor cost percentage *decreases*. Alternatively, lowering selling prices without increasing total revenue by an equal amount will result in *increases* in labor cost percentage.

Although labor cost percentage is easy to compute and widely used, it is difficult to express it as a measure of productivity over time. The reason: It depends on labor dollars spent and the sales dollars received for its computation.

Even in relatively noninflationary times, wages do increase, and menu prices are adjusted. Both activities directly affect the labor cost percentage, but not worker productivity. In addition, institutional foodservice settings, which often have no daily dollar sales figures to report, can find that it is not easy to measure labor productivity using labor cost percentages. Operators generally calculate and report guest counts or number of meals served rather than sales dollars earned.

Figure 10.7 summarizes key characteristics of the three measures of labor productivity just presented.

Regardless of the productivity measure used, if an operator finds labor costs are too high relative to sales produced, problem areas must be identified, and corrective action(s) must be taken. If the overall productivity of employees cannot be improved, other action(s) are important.

Measurement	Advantages	Disadvantages
Sales per Labor Hour = $\dfrac{\text{Total Sales}}{\text{Labor Hours Used}}$	1) Easy to compute 2) Does not vary with changes in the price of labor	1) Ignores price per hour paid for labor 2) Varies with changes in menu selling price
Labor Dollars per Guest Served = $\dfrac{\text{Cost of Labor}}{\text{Guests Served}}$	1) Easy to compute 2) Does not vary with changes in menu selling price 3) Can be used by nonrevenue-generating units	1) Ignores average sales per guest and, therefore, total sales 2) Varies with changes in the price of labor
Labor Cost % = $\dfrac{\text{Cost of Labor}}{\text{Total Sales}}$	1) Easy to compute 2) Most widely used	1) Hides highs and lows 2) Varies with changes in the price of labor 3) Varies with changes in menu selling price

Figure 10.7 Productivity Measures Summary

Another often ignored tactic that can increase employee productivity and reduce labor-related expenses is **employee empowerment**. This concept results from a decision by management to fully involve employees in the decision-making process in situations related to guests and employees.

Many experienced managers remember that it was once customary for management to (a) make all decisions regarding every facet of the operational aspects of its organization and (b) present them to employees as inescapable facts to be accomplished. Instead, an alternative approach occurs when employees are given the "power" to get involved.

Employees can be empowered to make critical decisions concerning themselves and, most importantly, the operation's guests. Many front-of-house employees work closely with guests, and numerous problems are more easily solved when employees are given the power to make it "right" for guests. Successful operators often find that well-planned and consistently delivered training programs can be helpful. Empowered employees can also yield a loyal and committed workforce that is more productive, is supportive of management, and will "go the extra mile" for guests. Doing so helps reduce labor-related costs, builds repeat sales, and increases profits.

Key Term

Empowerment (employee): An operating philosophy that emphasizes the importance of allowing employees to make independent guest-related decisions and to act on them.

Find Out More

One important challenge faced by many foodservice operators is that of recruiting and retaining qualified staff members.

The costs incurred in advertising vacant positions, interviewing potential staff members, and training those who are selected can be significant. These costs can be lessened when an operation provides a high-quality work environment that includes fair treatment of workers and the demonstration of concern for their well-being.

There are numerous strategies and tactics that can be implemented to develop a productive workforce. One good source of information to optimize the quality of the workforce while minimizing development costs is a useful book developed specifically for the foodservice industry.

To review the contents of this valuable employee recruitment and retention-related resource, part of the Wiley "Essentials" series for foodservice operators, go to www.Wiley.com., or "Amazon.com.," enter "Managing Employees in Foodservice Operations" and view the results.

What Would You Do? 10.1

Shaheed Kumar is the kitchen manager at the Talaron Corporation's International Headquarters. The facility he helps manage serves 3,000 employees per day. Shaheed needs an additional dishwasher as soon as possible. Despite advertising his vacant position for several weeks, Shaheed has had few applicants. He is now interviewing Donny, who is an excellent candidate with five years of experience who is now washing dishes at the nearby Downriver Rodeo Steakhouse.

Shaheed normally starts his new dishwashers at $14.00 per hour. Donny states that he currently makes $16.25 per hour; a rate that is higher than all but one of Shaheed's current dishwashers, many of whom have as much experience as Donny.

As they end the interview, Donny states that, while he would very much like to work at Talaron, he simply will not leave his current job if it means he must take a "pay cut."

Assume you were Shaheed. Would you hire Donny at a pay rate higher than most of your current employees? If so, what would you say to your current dishwashing employees if in the future Donny shared his pay information with them? If not, what would be your plan for filling your vacant dishwasher position?

The Importance of Controlling Other Operating Expenses

The control of food and beverage product costs and labor costs are extremely important in a foodservice operator's overall cost control efforts. There are, however, other expenses that significantly affect the success and profitability of a food-service operation.

These other expenses can account for a significant amount of the total cost of operating a foodservice business, and they include a variety of items. Some common examples include advertising, website maintenance, utility bills, equipment repair, and liability insurance. Other examples include property taxes and mortgage payments (if a building is owned by the operator) or rent/lease payments (if the building is owned by others).

Some of these other expenses are directly under management control. When following the Uniform System of Accounts for Restaurants (USAR), these types of expenses are reported on an operation's income statement as "Other Controllable Expenses" (see Chapter 2). Some other expenses, however, are not directly controllable by management and, as a result, they are reported on the income statement as "Noncontrollable Expenses."

Managing a foodservice operation's other expenses is just as important to its success as controlling its food, beverage, and payroll expenses. In most cases, the profit margins in many foodservice operations are small, so the control of all costs is critically important. Even in those situations that are traditionally considered nonprofit such as hospitals and educational institutions, all operating costs must be controlled. The reason: Dollars that are wasted in foodservice will not be available for use in other important areas of the facility.

Successful foodservice operators continually look for ways to control all operating expenses. However, sometimes the environment in which a business operates influences some operating costs in positive or in negative ways.

One good example relates to energy conservation and waste recycling. Energy costs are one of the other expenses examined in this chapter. In the past, a table service restaurant serving water to each guest upon arrival was a standard operating procedure (SOP). However, the rising cost of energy and an increased awareness about the environment of wasted resources have generated a new policy: serve water only upon request of arriving guests.

In most cases, guests have found this change acceptable, and the expense savings related to glass washing, equipment usage, energy, cleaning supplies, and labor costs can be significant. Note: In a similar manner, many operators today are finding that recycling fats and oils, cans, jars, and paper products can be good for the environment and for the property's bottom lines. Recycling these items reduces the cost of routine trash disposal and, in some communities, the recycled materials themselves have a cash value.

As they assess their operating costs, it is most often helpful for foodservice operators to consider various other expenses in terms of the costs (expenses) being either **controllable** or noncontrollable (see Chapter 2).

To see why, consider the case of Sven Hedin, the owner of a neighborhood tavern. Most of

Key Term

Controllable (expenses): Costs that can be directly affected by management's actions. Examples include labor and advertising costs.

Sven's sales revenue comes from the sale of beer, sandwiches, and his special pizza. Sven is free to decide on the monthly amount he will spend on advertising. Advertising expense, then, is under Sven's direct control and is considered a controllable other expense.

Some of his other expenses, however, are not under his direct control. One example is the license needed to operate his facility legally. The state in which Sven operates charges a license fee to all those businesses that sell alcoholic beverages. If his state increases the liquor license fee, Sven is required to pay the additional fee. In this situation, the alcoholic beverage license fee is a noncontrollable expense that is an expense outside of Sven's direct control.

As an additional example, assume an operator's business was part of a franchised quick-service chain that sells chicken sandwiches and chicken strips. Each month, the operator is assessed a $1,000 advertising and promotion fee by the chain's regional headquarters. The $1,000 is used to purchase television advertising time for the chain. This $1,000 charge is a noncontrollable operating expense because it is set by the franchisor and must be paid each month regardless of the revenue level the operator achieves.

Simply stated, a controllable expense is one in which decisions made by a foodservice operator can have the effect of either increasing or reducing the expense. A noncontrollable expense is one that the operator can neither readily increase nor decrease. Operators have some control over controllable expenses, but they usually have little or no control over noncontrollable expenses. As a result, operators should focus most of their attention on controllable (not noncontrollable) expenses.

It is also important to recognize that the items categorized as other expenses can constitute almost anything in the foodservice business. If a restaurant is a floating ship, periodically scraping the barnacles off the boat is categorized as another expense! If an operator serves food to oil field workers in Alaska, heating fuel for the dining rooms and kitchen will be another expense, and probably a very large one! If a company has been selected to cater an outdoor wedding, tent rental and tent erection costs may be significant other expenses.

Each foodservice operation will have its own unique list of required other expenses. The USAR lists several hundred other expense categories commonly incurred by foodservice operations. It is not possible, therefore, to list all imaginable expenses that could be incurred by every foodservice operator. The expenses that are incurred, however, should be recorded and reported in a meaningful way. For example, some restaurants incur napkin, straw, paper cup, and plastic lid costs. All these costs might be reported under a general grouping or cost category such as "Paper supplies and packaging."

Similarly, nearly all bars will incur expense for items such as jiggers, stir sticks, paper coasters, tiny plastic swords (for drink garnishes), and small paper umbrellas. These costs might be combined and reported in a cost category such as "Bar utensils and supplies."

When other expense cost groupings are used, they should make sense to the operator and be specific enough to let the operator know exactly what items are in the category. Although some operators prefer to make up their own other expense categories, the cost categories suggested in this book are those recommended by the USAR. The individual other expense cost categories recommended by the USAR are divided into two primary groups:

1) Other Controllable Expenses
2) Noncontrollable Expenses

Within each of these two major other expense groups, recommended groupings for individual cost categories are:

1) Other Controllable Expenses
   - Direct Operating Expenses
   - Music and Entertainment
   - Marketing
   - Utilities
   - General and Administrative Expenses
   - Repairs and Maintenance
2) Noncontrollable Expenses
   - Occupancy Costs
   - Equipment Rental
   - Depreciation and Amortization
   - Corporate Overhead (multi-unit restaurants)
   - Management Fees
   - Interest Expense

Within each of the two primary other expense cost subcategories, an operation's specific other expenses can be even further detailed to assist operators in their reporting and decision-making activities. For example, the USAR suggests the following individual costs be reported under the "Other Controllable Expenses" subcategory of "Utilities:"

✓ Electricity
✓ Gas
✓ Heating Oil and Other Fuel
✓ Recycling Expenses and Credits
✓ Trash Removal
✓ Water and Sewage

Similarly, those expenses that are noncontrollable may be detailed within their individual cost categories. For example, the USAR suggests the following individual costs be reported under the "Noncontrollable Expenses" subcategory of "Occupancy Costs" (see Chapter 1):

✓ Rent—Minimum or Fixed Amount
✓ Rent—Percentage Rent Amount
✓ Rent—Parking Spaces
✓ Ground Rent
✓ Common Area Maintenance
✓ Insurance on Building and Contents
✓ Other Municipal Taxes
✓ Personal Property Taxes
✓ Real Estate Taxes

One major purpose of the USAR is to provide operators and owners with guidance on how to best report the individual other expenses their businesses incur. The reasons why this is important will become very clear when operators analyze their cost control efforts using the income statement (see Chapter 12).

Find Out More

According to the U.S. Department of Energy, restaurants use about five to seven times more energy per square foot than other commercial buildings. High-volume quick-service restaurants (QSRs) may even use up to 10 times more energy per square foot than other commercial buildings.*

ENERGY STAR® is a joint program of the Environmental Protection Agency (EPA) and the Department of Energy (DOE). Its goal is to help consumers, businesses, and industry save money and protect the environment by adopting energy-efficient products and practices. The ENERGY STAR label identifies top-performing, cost-effective products, homes, and buildings.

The Department of Energy has also produced the *ENERGY STAR Guide for Cafés, Restaurants, and Institutional Kitchens* to help operators identify ways to save energy and water in businesses, boost their bottom lines, and help protect the environment. This resource also contains tips on how to upgrade equipment and highlights best practices that can positively impact a business's daily operations. This guide specifically addresses ENERGY STAR for Commercial Foodservice and other energy-saving options including Lamps and Lighting Fixtures, Heating, Ventilating and Air Conditioning (HVAC), and Water and Waste Management.

To learn more about money-saving energy-related tips, enter *"ENERGY STAR Guide for Cafés, Restaurants, and Institutional Kitchens"* in your favorite search engine and review the results.

*https://www.energystar.gov/buildings/resources_audience/small_biz/restaurants retrieved March 30, 2023.

Calculating Other Expense Costs

When assessing and managing other expenses, two control and monitoring calculations are very helpful to foodservice operators. These are:

1) Other expense cost percentage
2) Other expense cost per guest

Each of the above tools can be used effectively in specific management situations, so it is important to understand and know how to use both calculations.

Other Expense Cost Percentage

Other expenses can be analyzed based on their proportion (percentage) of total sales. The formula used to calculate a foodservice operation's other expense cost percentage is:

$$\frac{\text{Other Expense}}{\text{Total Sales}} = \text{Other Expense Cost \%}$$

The other expense cost percentage is a ratio that can be useful for budgeting and comparison purposes. For example, if an operation with an advertising expense of $5,000 in a month and achieved total sales of $78,000 for that same month, its advertising cost percentage would be calculated as:

$$\frac{\$5,000}{\$78,000} = 0.064 \text{ or } 6.4\%$$

The calculation required to calculate other expense cost percentage requires that the other expense cost category be divided by total sales. In many cases, this approach yields useful information. In some cases, however, this computation alone may not provide adequate information. Then using the concept of other expense cost per guest can be very helpful.

Other Expense Cost per Guest

Some foodservice operators find it helpful to calculate their other expenses on a cost per guest basis. The formula used to calculate a foodservice operation's other expense cost per guest is:

$$\frac{\text{Other Expense}}{\text{Number of Guests Served}} = \text{Other Expense Cost per Guest}$$

For example, if an operation with an advertising expense of $5,000 in a month served 10,000 guests for that same month, its advertising cost per-guest would be calculated as:

$$\frac{\$5,000}{10,000} = \$0.50 \text{ cost per guest}$$

In a variety of situations, an operator's other expense cost per guest provides more useful information than other expense cost percentage.

The other expense cost per guest formula can be used any time management believes it can be helpful, or when lack of a sales figure makes the regular computation of other expense cost percentage inappropriate.

In all cases, the effective control of production and service costs, labor costs, and the other expense costs described in this chapter are critical to the success of all foodservice operations. It is also important that food service operators have effective controls in place to protect the revenue they generate from their sales. The protection of a food service operator's revenue and other assets can make the difference between success and failure. The protection of a food service operations sales and other assets from internal and external threats is so important it will be the sole topic of the next chapter.

What Would You Do? 10.2

"Wow, this is ridiculous. I know it was hot last month, but I never expected our electric bill to look like this!" said Felix Unger, the manager of the Italian Garden restaurant.

Felix was talking to Myron Bollinger, the restaurant's owner.

Each month Felix and Myron reviewed the operation's P&L. The previous month had been a busy one, but the weather was unseasonably hot. As a result, the operation's electric bill was nearly 40% higher than the prior year. It was also much higher than the amount that was budgeted for the month's utility bill.

Since the utility bill was so high, the restaurant's other expense percentage for utilities had cut into the operation's profit in a significant way.

"Well, it certainly was hot last month," said Myron. "And our cooling costs were way up, but what can we really do about it?"

Assume you were Felix. Where would you go to find information about reducing HVAC costs in particularly high-cost months? What specific steps might you take in the future to help reduce energy costs in your operation during very warm periods, and how would you best communicate these steps to your staff?

Key Terms

Salaried employee	Ratio	Labor cost percentage
Exempt employee	Sales per labor hour	Empowerment
Job description	Labor dollars per	(employee)
Job specification	guest served	Controllable (expense)
Productivity (worker)		

Operator's 10-Point Tactics for Success Checklist

Evaluate your need for, and the current status of, each of the following operational tactics. For those tactics you think are important, but not yet in place, develop an action plan for its implementation including who will be responsible for the tactic's completion and the target date by which it should be completed.

Tactic	Don't Agree (Not Done)	Agree (Done)	Agree (Not Done)	If Not Done — Who Is Responsible?	If Not Done — Target Completion Date
1) Operator understands the relationship between profits and controlling labor costs.	____	____	____		
2) Operator can state the importance of recognizing key nonpayroll payment factors that directly impact total labor costs.	____	____	____		
3) Operator knows how to calculate and assess the "Sales per Labor Dollar" measure of productivity.	____	____	____		
4) Operator knows how to calculate and assess the "Labor Dollars per Guest Served" measure of productivity.	____	____	____		
5) Operator knows how to calculate and assess a "Labor Cost Percentage."	____	____	____		
6) Operator recognizes the impact employee empowerment can have on reducing labor costs and improving profits.	____	____	____		
7) Operator understands profits can be increased by properly controlling "other" operating expenses.	____	____	____		

Tactic	Don't Agree (Not Done)	Agree (Done)	Agree (Not Done)	Who Is Responsible?	Target Completion Date
				If Not Done	
8) Operator can state the difference between a controllable and a noncontrollable other operating expense.	____	____	____		
9) Operator can calculate other expense cost as a percentage of total sales.	____	____	____		
10) Operator can calculate other expenses on a cost per guest basis.	____	____	____		

11

Revenue Control

What You Will Learn

1) How to Identify External and Internal Threats to Revenue
2) How to Create Countermeasures to Address Revenue Theft
3) How to Establish and Monitor an Effective Revenue and Asset Security System

Operator's Brief

One of a foodservice operator's most important responsibilities is to protect business assets including cash and products from theft and fraud. Foodservice operations can be an easy victim of these crimes, and control systems maintaining cash and other asset securities are essential.

In this chapter, you will learn about effective revenue control systems and how their components are developed, implemented, and monitored. A foodservice operation's cash assets are generated by sales to guests. Unfortunately, cash can be tempting to dishonest persons, and its security is subject to external and internal threats that must be controlled.

In this chapter, you will learn important procedures for safeguarding cash from when it is received from guests until it is deposited in an operation's bank account. There are several objectives and useful strategies important as revenue security programs are developed, and this chapter explains them.

As revenue security is implemented and monitored, operators must pay special attention to five key areas. Each is detailed in this chapter, and you will learn why it is necessary to properly verify your operation's:

1) Receivable Revenue
2) Total Guest Charges
3) Sales Receipts
4) Sales (Bank) Deposits
5) Accounts Payable (AP)

The Importance of Revenue Control

Foodservice operators need an effective system to control cash revenues. These controls are important because many foodservice operations have hundreds (or more!) of daily cash sales transactions, and more than one employee typically handles the cash.

While some foodservice operators consider revenue generated to be the same as cash, it is not. For example, if a foodservice operator makes a sale to a guest who uses a credit card for payment, that operator will have generated revenue, but will not have cash until the credit card company deposits the cash in the operator's bank account.

Similarly, if the foodservice operation is hosting a future wedding event, it may require a cash deposit in advance. In this case the operation will have received cash, but it will not actually have generated any revenue until the event is held. Revenue is generated when products or service are delivered, and cash is generated when money is received. In this chapter, the term "revenue" will be used to refer to the sales achieved by a foodservice operation, regardless of the timing of cash payments made to it.

All food service operators must take appropriate steps to protect their revenue. **Fraud** and **embezzlement** can occur frequently in some foodservice operations. Fortunately, there are several ways operators can reduce opportunities for these thefts to occur.

Key Term

Fraud: Deceitful conduct to manipulate someone into giving up something of value by (a) presenting as true something known by the fraudulent party to be false or (b) by concealing a fact from someone that may have saved him or her from being cheated.

Key Term

Embezzlement: A crime in which a person or entity intentionally misappropriates assets that were entrusted to the person or entity.

In most cases, there are three primary factors necessary for employee fraud and/or embezzlement to occur:

1) Need: Economic or psychological motives can encourage employees to steal. Some operators attempt to study their employees' lifestyles to determine whether selected staff members may steal. However, a better approach is to develop systems that reduce theft opportunities and make it difficult for employees to "try to beat the system."
2) Failure of conscience: Sometimes thieves rationalize stealing to justify taking someone else's property. For example, some foodservice staff might believe they deserve more money because they are paid too little, and "stealing a little is okay because the business can afford it."
3) Opportunity: Some otherwise honest employees may be tempted to commit theft when they are given the opportunity to do so. Foodservice operators can discourage theft by implementing systems and procedures that minimize opportunities for employee theft.

Of the three factors noted above, foodservice operators have the most direct influence over theft opportunities. Therefore, cash and revenue control procedures should be designed to minimize these theft possibilities.

Most foodservice operations have some general operating characteristics that make them more vulnerable to theft than many other businesses. These include:

✓ Large numbers of individual cash transactions
✓ Items of relatively high value commonly used or available
✓ Workplace availability of products employees must otherwise purchase

Increasingly, foodservice guests use electronic payment forms rather than cash when paying their bills. However, cash is often exchanged between guests and employees. As a result, **cash banks** made available to cashiers and servers can create potential theft risks as well.

In some operations, the use of relatively unskilled employees working in low-paying positions can yield high employee turnover rates.

Key Term

Cash bank: The currency and coin holder used to store money received from guests for their cash purchases and to provide change for cash purchases.

These, in turn, can influence the internal control environment. Many foodservice operations are small, and even large operations may be organized into several small revenue outlets with individual dining rooms, multiple bars, drive-thrus and, perhaps, take-out stations. These multiple sales locations can create even stronger needs for effective internal control systems that address both external and internal threats to revenue security.

External Threats to Revenue Security

Foodservice guests can be threats to an operation's revenue, and an operation can lose revenue because some guests wanted to defraud it. This activity can take a variety of forms including that guests **walk** (skip) without paying their bill. This type of theft is less likely in a quick-service restaurant (QSR) or fast casual operation because pay-

Key Term

Walk (bill): Guest theft that occurs when a guest consumes products that were ordered and leaves without paying the bill. Also referred to as a "skip."

ment is typically collected before, or at the same time, guests receive their menu items.

In cases where a guest is in a busy table-service restaurant's dining room, it is easier for one or more persons in a dining party to leave without paying while their server is busy with other guests. In fact, it is sometimes easy for a guest or entire party to leave without settling their bill unless all staff members are vigilant. To help reduce this type of guest theft, implementation of the steps in Figure 11.1 can be helpful.

---

1) If guests order and consume their food before payment, servers can present the bill promptly when guests finish eating.
2) If the operation has a cashier in a central location in the dining area, they should be available and visible at all times.
3) If each server collects their own guest's charges, they should return to the table promptly after presenting the guest's bill for payment.
4) Train employees to observe all exit doors near restrooms or other areas of the facility that may provide opportunities to exit dining areas without being easily seen.
5) If an employee sees a guest attempting to leave without paying the bill, they should notify the manager or person in charge immediately.
6) When approaching someone who has left without paying the bill, the manager should ask if the guest has inadvertently "forgotten" to pay. (Note: In most cases, the guest will then pay the bill.)
7) If a guest still refuses to pay or flees the scene, the operator should make a written note of the following:
   a) Number of guests involved
   b) Amount of bill
   c) Physical description of guest(s)
   d) Vehicle description, if applicable, with license plate number when possible
   e) Time and date of the incident
   f) Name of the server(s) who served the guest(s)
8) If the guest is successful in fleeing the scene, police should be notified. Note: In no case should staff members, supervisors, or operators attempt to physically detain the fleeing guest(s). The reason: Liability is involved if an employee or guest is hurt, and this may result in a far greater loss than the value of a skipped food and beverage bill.

Figure 11.1   Steps to Reduce Guest Walks (Skips)

Another form of external theft can result from the successful efforts of a **quick-change artist**. For example, a quick-change artist who should have received $5 in change may use a confusing routine to secure $15. To prevent this from happening, operators must train cashiers and other employees who accept cash to notify management immediately if there is suspicion of attempted fraud from quick-change routines.

Key Term

Quick-change artist: An individual who attempts to confuse a cashier into giving excessive change.

Internal Threats to Revenue Security

Most foodservice employees are honest, but some are not. In addition to protecting revenue from unscrupulous guests, operators must also be aware of employees who attempt to steal revenue from their properties.

Cash is the most readily usable asset in a foodservice operation, and it is often a major target for dishonest employees. In general, theft by service personnel does not occur by an employee removing large sums of cash at one time because this method would normally be relatively easy for operators to detect. Rather, dishonest service personnel may use numerous methods of removing a small amount of money at different times.

One common theft technique used by servers involves failure to record a guest's order in the operation's point-of-sale (POS) system. Instead, the server charges the guest and keeps the revenue from the sale. To reduce this possibility, operators must insist that (a) all sales be recorded in the POS system and (b) match products sold to revenue received. (Note: This type of checking is often done as part of the cash-handling employee's end-of-shift check-out procedures.)

The POS system assigns a unique transaction number to the sale just as numbered guest checks did before the introduction of POS systems. A sale's transaction number (or a numbered guest check) is an electronic or handwritten record of what the guest purchased and how much the guest was charged for those item(s).

The use of electronic guest checks produced by the POS system is standard in the foodservice industry. A rule for all food and beverage production personnel is that no food or beverage item should be issued to a server unless the server first records the sale in the POS system. The items ordered are then printed out or displayed in the operation's production area.

Increasingly, this method of entering guest orders can be accomplished in one step instead of two. Handheld and wireless at-the-table order-entry devices now allow servers (and guests in some cases!) to enter orders directly into an operation's POS system. This direct data entry system is fast, and it eliminates mistakes made when transferring handwritten guest orders to the POS system.

Regardless of how the order is created, kitchen and bar personnel should not issue any products to the server without this uniquely numbered sales transaction. When the guest wants to leave the facility, the cashier (or server) retrieves the transaction and prepares a bill for payment by the guest. The bill includes the charges for all items ordered by guests plus any service charges and taxes due, and the guest then pays the bill.

Another method of employee theft involves entering sales but failing to collect payment. To prevent this theft alternative, management must have systems in place to identify **open checks** during and after each server's work shift.

Unless all open checks are ultimately presented to guests for payment and closed out, the value of menu items issued will not equal the money collected for those items' sale.

The totals of all transactions entered in the POS system during a predetermined time are electronically tallied, and operators can compare sales recorded by the POS system with the money (including charged debit and/or credit cards and gift cards) in the cash drawer.

For example, a cashier working a shift from 7:00 a.m. to 3:00 p.m. might have recorded $2,000 in sales (including taxes) during that time. If that were the case and if no errors in handling change occurred, the cash drawer should contain the $2,000 in sales revenue (in addition to the amount in the drawer at the beginning of the shift). If the drawer contains less than $2,000, it is said to be **short**; if it contains more than $2,000, it is said to be **over**.

Cashiers rarely steal large sums of cash from the cash drawer because this type of theft is easily detected. Operators should implement a policy to ensure any significant cash shortages or overages will be investigated. (Note: Some operators believe that only cash shortages, not overages, need to be monitored, but this is not true.)

Key Term

Open check: A guest check that initially authorizes product issues from the kitchen or bar for which collection of the amount due has not been made (the account is not closed). Therefore, the amount due has not been added to the operation's sales total.

Key Term

Short (cash bank): A cashier's bank that contains less than the amount of cash, credit, or debit card, or other authorized payment form based on the number of menu items sold.

Key Term

Over (cash bank): A cashier's bank that contains more than the amount of cash, credit, or debit card, or other authorized payment form based on the number of menu items sold.

Consistent cash shortages may be an indication of employee theft or carelessness and must be investigated. Cash overages, too, may result from sophisticated theft by the cashier. For example, assume a cashier defrauds an operation by removing and keeping $18.00 from a cash drawer but falsely reduces actual sales records by $20.00. The result is a $2.00 cash "overage!"

Even if operators implement controls to make internal theft difficult, the possibility of significant fraud still exists. Consequently, some foodservice operators protect themselves from employee dishonesty by **bonding** their employees. Bonding involves purchasing an insurance policy against the possibility that employee(s) will steal.

When bonded, an employer can be covered for the loss of money or other property sustained through dishonest acts. Bonding can cover many acts including larceny, theft, embezzlement, forgery, misappropriation, or other fraudulent or dishonest acts committed by an employee alone or in **collusion** with others. Essentially, a business can select from several bonding options:

Key Term

Bonding (employee):
Protection from loss resulting from a criminal event involving employee dishonesty and theft such as loss of money, securities, and other property.

Key Term

Collusion: Secret cooperation between two or more individuals that is intended to defraud an operation.

1) Individual—covers one employee (e.g. a foodservice operation's bookkeeper)
2) Position—covers all employees in a specific position (e.g. all servers and/or all cashiers)
3) Blanket—covers all employees

If an employee has been bonded and an operator can determine that the employee was involved in the theft of a specific amount of money, the business is reimbursed for all or part of the loss by the bonding company.

Although bonding will not eliminate all theft, it is a relatively inexpensive way to help ensure that an operation is protected from theft by employees who handle cash or other forms of operating revenue. Note: The bonding company will likely require detailed background information on employees before bonding them, and this is also an excellent pre-employment check to verify an employee's "track record" in prior jobs.

Operators must recognize that even very good revenue control systems present the opportunity for theft if management is not vigilant, and this is especially so if two or more service employees work together to defraud the operation. Figure 11.2 identifies some common methods of internal theft involving service employees in a foodservice operation.

The scenarios addressed in Figure 11.2 do not include all possible methods of revenue loss. However, successful operators use an effective revenue security system to ensure all menu items sold generate sales revenue for the property.

1) Omits recording the guest's order and keeps the money the guest pays
2) Voids a sale in the POS system but keeps the money the guest paid
3) Enters another server's password in the POS system and keeps the money
4) Fails to finalize a sale (keeps a check open) and keeps the money
5) Charges guests for items not purchased and then keeps the overcharge
6) Changes the totals on payment card charges after the guest has left
7) Enters additional payment card charges and pockets the cash difference
8) Incorrectly adds legitimate charges to create a higher-than-appropriate total with the intent of keeping the overcharge
9) Purposely shortchanges guests when giving back change with the intent of keeping the extra change
10) Charges higher-than-authorized prices for products or services, records the proper price, and keeps the overcharge
11) Adds a coupon to the cash drawer and, at the same time, removes sales revenue equal to the coupon's value
12) Declares a transaction to be complimentary (comped) after the guest has paid the bill
13) Engages in collusion to defraud the operation

Figure 11.2   Common Methods of Revenue Theft by Service Employees

---

Find Out More

Dishonest foodservice employees can steal from their employers, but they also can also steal from an operation's guests.

In many cases, foodservice customers paying their bills hand over their credit cards to their servers, but they may not see where their card goes. This leaves opportunities for a guest's credit card information to be stolen on a credit-card skimming device that reads the magnetic stripe on a credit or debit card. When inserted into a card reader by a dishonest employee, the employee stores the card number, expiration date, and cardholder's name for later illegal use by the dishonest employee. Alternatively, an employee may also use a cell phone to photograph the front and back of a guest's credit card to obtain the name, credit card number, and the 3- or 4-digit security code number found on the back of the card.

Foodservice operators have a responsibility to help ensure the security of their guest's payment data. Operators should remain updated about methods that could defraud their guests. To learn more about what can be done to help ensure the security of each guest's payment information, enter "preventing credit card theft by restaurant employees" in your favorite search engine and view the results.

What Would You Do? 11.1

"The beauty of our system," said Ned Carson, "is that you can monitor the actions of all your employees and supervisors."

Ned was talking to Jada Tuller, the owner of Fazziano's Italian Kitchen. Jada had called POS-Video Security, the company Ned represented because, for the second time this year, Jada had discovered a case of employee/supervisor collusion. Working together, the employee and supervisor stole revenue from their restaurant by manipulating their unit's POS system.

"I'm pretty sure I understand your system, but please go over it one more time," said Jada.

"Okay, Jada," said Ned. "Essentially, our new system goes beyond traditional viewing methods by coordinating the video being recorded with the data mined from your POS system to create detailed, customized video reports. Potentially fraudulent activity including manager overrides, coupons, or comps, or even a cash drawer being open for too long, is tracked, and the corresponding video review can be searched by transaction number. Data reports and streaming video, both real time and stored, can be accessed securely and remotely on a PC, smart device, or in the cloud."

"So, for example," replied Jada, "when a sales void occurs, your system identifies the portion of video that was recording at the time of the void and then allows me to view just that portion of the video so I can see what was happening in the restaurant during the transaction."

"Exactly," said Ned.

Assume you were Jada. What types of employee fraud do you think could be uncovered using the technology offered by POS-Video Security's new product? Do you think the behavior of some dishonest cashiers and supervisors would change if they knew their actions were being video recorded? Explain your answer.

Developing a Revenue Security Program

Chapter 1 of this book introduced a foodservice operator's basic profit formula:

Revenue − Expenses = Profit

A close examination of the formula might lead some operators to think that 50% of their time should address managing and protecting revenue, and 50% of their time should consider managing expenses. This may be reasonable because all cost control systems will be of little use if operators cannot initially collect the revenue generated by their businesses, deposit the revenue into their bank accounts, and spend it only for legitimate expenses.

Errors in revenue collection or other asset security issues can result from simple employee mistakes or, sometimes, outright theft by guests and/or employees. As a result, an important part of every operator's cost control-related job is to devise revenue control systems to protect income. This is true regardless of whether the revenue is cash, checks, credit or debit card receipts, coupons, meal cards, or another guest payment method.

Objectives of Internal Revenue Control Systems

The American Institute of Certified Public Accountants (AICPA) has a longstanding definition of internal control:

> *Internal control comprises the plan of organization and all of the coordinate methods and measures adopted within a business to safeguard its assets, check the accuracy and reliability of its accounting data, promote operational efficiency, and encourage adherence to prescribed managerial policies.*[1]

This definition indicates that internal control relies on an organization plan and use of methods and measures to attain four key objectives:

1) Safeguard assets
2) Check accuracy/reliability of accounting data
3) Promote operational efficiency
4) Encourage adherence to prescribed managerial policies

Each of these objectives are addressed when foodservice operators develop and implement effective revenue control systems.

Elements of Internal Revenue Control Systems

Regardless of the specific foodservice operation, all effective revenue control systems have similar elements.

Proper Leadership

Leadership is critical to the success of a foodservice operation's internal control system. Effective policies must be developed, clearly communicated, and consistently enforced. Each management level must be responsible for ensuring that applicable control procedures are adequate, and that any exceptions to policies are minimized and justified.

---

1 American Institute of Certified Public Accountants, Committee on Auditing Procedures, Internal Control-Elements of a Coordinated System and Its Importance to Management and the Independent Public Accountant (New York, 1949).

Assigned Responsibility

When possible, responsibilities for a specific control activity should be assigned to one individual. That staff member can be given a set of standard operating procedures with the expectation they will be followed. When responsibility is given to one person, an operator knows where to start looking if a problem is identified.

For example, a cashier should be solely and fully responsible for a specific cash bank. To ensure this, no one except that employee should have access to the bank during the cashier's shift. In addition, there should be no sharing of the cash bank and/or responsibility for it.

Separation of Duties

Separation of duties occurs when different personnel are assigned to accounting, asset responsibility, and production activities. Duties within the accounting function should also be separated. For example, different personnel should maintain an operation's **general ledger** and the operation's **cash receipts journal**.

The major objective of separating duties is to prevent and detect errors and theft. Unfortunately, it is not always possible to separate duties in very small foodservice operations. In these situations, operators must often assume multiple duties.

Key Term

General ledger: The main or primary accounting record of a business.

Key Term

Cash receipts journal: An accounting record used to summarize transactions related to cash receipts generated by a foodservice operation.

Approval Procedures

Foodservice operators must authorize every business transaction. Authorization may be either general or specific. General authorization occurs when all employees comply with selected procedures and policies when performing their jobs. For example, in all operations, servers must sell food and beverage products at the prices listed on the menu.

Specific variations from this general policy and any other significant issues must be approved by management.

Proper Recordkeeping

The accurate recording of security-related information is essential for effective internal control. Asset security documents including purchase orders, inventory evaluation sheets, sales records, and payroll schedules should be designed to be easy to complete and understand.

Written Policies and Procedures

Employees can only be expected to follow policies they understand. Significant security-related policies and procedures should be in writing. They should also be included in employee manuals and new employee orientation programs for all employees. Written policy and procedure manuals help ensure that all employees are treated fairly and similarly if policy violations occur.

Physical Controls

Physical controls are often necessary to properly safeguard assets. Examples include security devices such as cash safes and locked refrigerated and dry storage areas.

Performance Checks

Performance checks help to ensure all elements in the internal control system are functioning properly. Whenever possible, the checks should be independent (i.e., the person doing an internal verification of the information should not be the same person responsible for collecting the data initially).

In large businesses, auditors who are independent of both operations and accounting personnel report directly to top-level management. In smaller operations, operators should conduct most, but not all, performance checks. For example, an operation's **bank reconciliation** should be done by personnel who are independent from those who initially account for cash receipts and vendor payments.

Key Term

Bank reconciliation: A process performed to ensure that a company's financial records are correct. This is done by comparing the company's internally recorded amounts with the amounts shown on its bank statement. Any differences must be justified and, when there are no unexplained differences, accounting personnel can affirm that the bank statement has been reconciled.

Implementing and Monitoring a Revenue Security Program

In its simplest form, revenue control and security involve matching products sold with money received. Implementing and monitoring a total revenue security program involves ensuring that systems and recordkeeping processes are in place to allow foodservice operators to always verify that:

1) Documented Menu Item and Bar Orders = **Receivable Revenue**

Key Term

Receivable revenue: The expected sales income when all food and beverage orders have been properly documented.

2) Receivable Revenue = Total Guest Charges
3) Total Guest Charges = Sales Receipts
4) Sales Receipts = Sales (Bank) Deposits
5) Sales (Bank) Deposits = Funds Available to Pay Accounts Payable (AP)

The term "receivable revenue" means that a specified amount (charges) must be included in Step 2 and actually paid for in Step 3.

To illustrate these five verification points required to ensure an effective revenue security program, consider Maoli Felize De la Cruz, who operates a Dominican Republic-themed restaurant in New York.

Maoli considers her restaurant to be a family-oriented establishment. It has a small (20 seat) cocktail area, and it seats 100 guests in the dining room. When she initially started the restaurant, she did not give much thought to the design of her revenue control systems because she generally worked in the operation every day. Due to her success, she now spends more time away from the property to develop a second restaurant, and she requires the security of an adequate revenue management system and the ability to review it quickly.

Maoli has begun to develop a revenue security system by concentrating her efforts on ensuring that the five steps noted above are always followed:

Receivable Revenue = Total Guest Charges = Sales Receipts = Sales (Bank)
Deposits = Funds Available to Pay Legitimate Accounts Payable (AP)

Verification of Receivable Revenue

A key to verification of receivable revenue in a revenue security system is to follow one basic rule: "*No menu item shall be issued from the kitchen or bar unless a permanent record of the item prepared has been made.*"

This means that the kitchen (and bar) should not fill any server request for menu items unless the product request has been documented in writing or electronically. Note: in some small restaurants, the server's written request for food or beverages takes the form of a single (or multi-copy) written guest check designed specifically for the purpose of revenue control. The top copy of this multi-copy form generally is sent to the kitchen or bar. The guest check, in this case, becomes the documented request for the food or beverage product ordered.

This "paper-only" system can work, but it is subject to many forms of abuse and fraud. Therefore, when possible, foodservice operations should utilize an electronic POS system in which the "guest check" consists of an electronic record of product requests and issues. Then a guest's order is viewed by the production staff on a computer screen, or, in other cases, the POS system prints a hard copy of the order that will be used by production staff filling the order.

In either case, the software within the POS system creates a permanent record of the transaction and issues a unique transaction number to identify the requested product(s). This record authorizes kitchen personnel to prepare food or the bartender to make a drink. If a foodservice operation elects to supply its employees with meals during work shifts, these meals should also be recorded in the POS system.

In the bar, the principle of verifying all receivable revenue is even more important. Bartenders should be instructed to never prepare a drink unless that drink has first been recorded in the POS system. This should be the standard operating procedure (SOP), even if the bartender is working alone. This rule regarding menu item ordering is important for two reasons. First, requiring a permanent documented order helps ensure that there is a record of each product's production. Second, this record can be used to verify both proper inventory usage and product sales totals.

Maoli enforces this basic rule by requiring that no item be served from her kitchen or the bar without the sale first being entered into her POS system. If her verification of product production (menu items produced) works correctly, Maoli will note the following formula is always correct:

Documented Menu Item and Bar Orders = Receivable Revenue

Technology at Work

A high-quality POS system always provides detailed reports. The sales report dashboard on a POS system provides an overview of all transactions completed during a selected time period. This includes net sales, service charges, tips, total guests served, table turn times, and a breakdown of all service types and payment methods. It should also provide detailed sales exception reports that allow users to quickly see an overview of all voids, discounts, and refunds. The reports also allow users to identify the specific servers and operators who are giving and approving sales exceptions such as voiding receipts, discounting food, and offering refunds.

It is important to recognize that, regardless of its sophistication, a POS system will not "bring" control to a foodservice operation. A high-quality POS system can, however, take good control systems that have been carefully designed by management and add to them speed, accuracy, and/or additional information.

Properly utilized, a POS system is of immense value. If, however, an operation has no formal revenue security plan, the POS system simply becomes a high-tech adding machine used primarily to sum guest purchases and little more. Properly selected and utilized, however, POS systems play a crucial role in the implementation of an operation's complete revenue security system.

To review the cash control-related features of some currently popular POS systems, enter "POS revenue control reports for restaurants" in your favorite search engine and view the results.

Experienced foodservice operators know that, despite their best efforts, it is possible for employees to issue menu items without a documented product request when:

1) Two or more employees work together to defraud the operation. Collusion of this type can be discovered when operators use a system to carefully count the number of items removed from inventory and then compare that number to the number of products produced and issued.

2) A single employee (example: a bartender working alone) is responsible for both making and filling the product request. If fraud occurs, operators can uncover it when they carefully compare the number of items (or, in some cases such as bottled beers and wine sold by the bottle) removed from inventory with the number of recorded product issues.

Verification of Guest Charges

An operation's production staff must prepare and distribute products only in response to a properly documented request. It is critical that it (the documented request) results in charges to the guest. The reason: It makes little sense to enforce verification of product production without also requiring the service staff to ensure guest charges match these requests.

When an operator insists that no products be issued without a POS-generated request, the managerial goal is to ensure that the menu items and/or beverage products equal guest check totals. In other words, all issued products should result in appropriate charges to the guest.

When properly implemented, this second step of the revenue control system will ensure the following formula always holds true:

Receivable Revenues = Total Guest Charges

Maoli has now implemented the first two key revenue control principles. The first is that no items can be sold from the kitchen or bar unless the production order is documented. The second principle is that all guest charges must match revenue receivables generated by the menu items and/or beverage items served.

With these two systems in place, Maoli can deal with many potential problems. If, for example, a guest has "walked" the check, the operation has a duplicate record of the transaction. The POS will have recorded the products sold to this guest, who sold them and, perhaps, information including time of the sale, number of guests, and total sales value of menu items produced.

The POS system Maoli uses also ensures that service personnel cannot "change" the prices charged for items sold. Note: This would likely be possible in an operation using manual (paper) guest checks.

To complete this aspect of her control system, Maoli implements a strict policy regarding the documentation of employee meals (a labor cost). Recall that this amount is needed to accurately compute cost of sales (see Chapter 2) when the property's income statement is prepared.

Maoli is now ready to address the next major component in her revenue security system: the actual collection of guest payments. The sum of these payments will represent Maoli's actual sales receipts.

Verification of Sales Receipts

The term "sales receipts" refers to the actual revenue received by the cashier, server, bartender, or other designated personnel in payment for products served and any payments made online. In Maoli's case, this means all sales revenue from her restaurant and bar.

The essential principle to recognize in this step is that two individuals (the cashier and member of Maili's management team) must verify sales receipts. Although this will not prevent possible collusion by two individuals, the sales receipt verification should be a two-person process. Maoli wants to ensure that the amount of cash collected when added to her non-cash (including credit and debit card) guest payments matches the dollar amount she charged her guests as recorded in the POS system.

In most operations, individual guest charges are recorded only in the POS system. This is the case, for example, in a QSR or cafeteria, where food purchases are totaled and paid for at the same time. In these instances, the POS system provides an immediate and accurate total of guest and other charges. Receipts collected should always equal these charges. If Maoli's revenue security system is working properly, the following formula noted above will always be true:

Total Guest Charges = Sales Receipts

Note that total POS-recorded charges will consist of all sales, service charges, tips, and guest-paid taxes, and they represent the total revenue (money) the operation should receive.

Verification of Sales Deposits

Most foodservice operations make a sales deposit each day the property is open because keeping excessive amounts of cash on site is not advisable. It is strongly recommended that only management make the actual bank deposit of daily sales receipts. A cashier or other clerical assistant may complete a deposit slip, but the operator should be responsible for monitoring the sales deposit.

This task involves the actual verification of the deposit's contents and the process of matching bank deposits with actual sales receipts. These two numbers should match. If Maoli, or a member of her management team deposits Thursday's sales on Friday, the Friday deposit should match the sales amount of Thursday. If it does not, her operation has experienced some loss of revenue.

It is this step of the revenue control system in which embezzlement is most likely to occur, and the crime can go undetected for a long time because the embezzler is often a trusted employee. Falsification of, or destroying, bank deposits is a common method of embezzlement.

To prevent this activity, Maoli should take the following steps to protect her sales deposits:

✓ Make bank deposits of cash and checks daily, if possible.
✓ Ensure the person preparing and making the deposit is bonded.
✓ Establish written policies for completing bank reconciliations: the regularly scheduled comparison of the business's deposit records with the bank's deposit records.
✓ Payment card fund transfers to a business's bank account should be reconciled each time they occur. Today, in most cases, cash and non-cash payment reconciliations can be accomplished daily using online banking features.
✓ Review and approve written bank statement reconciliations at least monthly.
✓ Change combinations on office safes periodically and share the combinations with the fewest possible employees.
✓ Require all cash-handling employees to take regular and uninterrupted vacations at least annually. This allows another employee to review and potentially uncover any improper practices.
✓ Consider employing an outside **auditor** to examine the accuracy of deposits on an annual basis.

If the verification of sales deposits is done correctly, and no embezzlement is occurring, the following formula should always hold true:

Sales Receipts = Sales (Bank) Deposits

Sales deposit records are maintained in an operation's **back office accounting system**,

Key Term

Auditor: An accounting professional who specializes in studying a foodservice operation's internal controls and analyzing its basic accounting system to ensure all financial information is properly recorded and reported.

Key Term

Back office accounting system: The accounting software used to maintain a business's accounting records that are not contained in its POS system. This typically includes items such as payroll records, accounts payable (AP) and accounts receivable (AR), taxes due and payable, net profit and loss summaries, and relevant balance sheet entries.

and this data must be regularly and carefully monitored by an operation's managers or owners.

Technology at Work

Modern POS systems are a critical tool in maintaining revenue control, but they are not the only helpful item. To ensure an effective revenue control system, foodservice operators must also have an up-to-date and effective back office accounting system.

Note: The term "back office" originated when early companies designed their offices, so the front portion was a workplace for the employees who interact with guests. The back portion of the office were the workplaces of associates with little guest interaction such as accounting clerks.

Back office accounting software is used by foodservice and other businesses to help track income and expenses, create invoices, calculate sales taxes due, price recipes and menus, and more.

The best back office accounting systems can be interfaced (electronically connected) to an operation's POS system. Doing so makes it easy to monitor both revenue collection and revenue deposits and cash account balances.

To learn more about the features included in these essential accounting systems, enter "back office accounting systems for restaurants" in your favorite search engine and review the results.

Verification of Accounts Payable (AP)

AP, as defined in this step refers to the legitimate amounts owed to suppliers or vendors for the purchase of products or services. The basic principle to be followed when verifying AP is:

> *"The individual authorizing the purchase should verify the legitimacy of the vendor's invoice before it is paid."*

Vendors' payments are often an overlooked potential threat to the security of a foodservice operation's revenue. Of course, an operation should pay all valid expenses. However, both external vendors and an operation's employees can attempt to defraud a foodservice operation by falsifying invoices.

For example, consider Maoli. She has just received an invoice for fluorescent light bulbs. The invoice is for over $400 dollars, but the invoice indicates that only two dozen bulbs were delivered.

Maoli is not familiar with this specific vendor, but the delivery slip included with the invoice was signed (six weeks ago!) by her receiving clerk. As a result, Maoli and her operation may be victims of a vendor's invoice scam that threatens her operation's revenue.

Find Out More

Foodservice operations are popular targets for invoice fraud. In these scams, criminals send bills for goods or services a business never ordered or received. The scam succeeds mainly because the invoices look legitimate, and unsuspecting accounts payable employees do not look closely to determine if they are real. They simply make the payment, often thinking someone else in the operation placed the order.

To protect your operation from invoice scams, you must:

1) Be cautious when processing invoices: Ensure your AP personnel are aware of these scams and are cautious when processing invoices.
2) Verify unfamiliar vendors and do not purchase from new suppliers or pay invoices from unfamiliar vendors until you verify their existence and reliability.
3) Check invoices against original purchase orders before paying them: a written purchase order or similar record should exist for each invoice an operator receives.

To learn more about additional procedures and policies operators can implement to minimize the chances of becoming a victim of invoice fraud, enter "protecting against invoice fraud in restaurants" in your favorite search engine and review the results.

Dishonest suppliers can take advantage of weaknesses in an organization's purchasing procedures and/or in unsuspecting employees who may not be aware of their fraudulent practices. In addition, the supplies delivered by these bogus firms are most often highly overpriced and of poor quality.

When her revenue security program is working properly, Maoli can confirm that:

Sales (Bank) Deposits = Funds Available to Pay Accounts Payable (AP)

Funds available for AP should only be used to pay legitimate expenses resulting from a purchase verified by authorized personnel within the foodservice operation.

When Maoli properly completes the building of her revenue control system, its key features can be summarized as shown in Figure 11.3.

1) No menu items shall be produced in the kitchen or bar unless a permanent record of the items' production is made.
2) Revenue receivables must equal total guest charges.
3) Both the cashier and the operator must verify all sales receipts.
4) Management must personally verify all bank deposits.
5) Management or the individual authorizing the purchase should verify that all vendor invoices represent legitimate AP expenses.

Figure 11.3 Revenue Control Points Summary

The best revenue control programs help foodservice operators maintain the security of their cash and other assets. Previous chapters of this book focused on optimizing product, service, labor costs, and other operating expenses. Information in those chapters allows foodservice operators to better understand what these costs *are*.

The next, and final, chapter of this book addresses budgeting for and analyzing an operation's actual costs to help ensure these costs are consistently what they *should be*.

What Would You Do? 11.2

Bart Stevens worked for 15 years as the head snack bar cashier for the Downtown Sports Arena Complex, a facility whose food concessions were managed by JeAnna Arbor's "Elite Catering" company.

Bart had twice won the company's "Employee of the Year" award, and JeAnna considered Bart to be a valued and trusted employee who, on many occasions, performed far above and beyond what was required.

JeAnna was very surprised when newly installed video surveillance equipment confirmed that Bart, despite strict written rules against it, had recently given free food and beverages to friends who visited the arena.

When confronted with the video evidence, Bart admitted the conduct, apologized profusely, and asked JeAnna for a second chance. He promised never to give free food or beverages to anyone in the future.

On the advice of the company's attorney, JeAnna is documenting, in writing, her decision on handling the situation.

Assume you were JeAnna. Do you believe an employee caught defrauding their employer should ever be given a second chance? If so, under what circumstances? What impact will your decision in this case have if, in the future, other employees are caught stealing from your operation?

Key Terms

Fraud
Embezzlement
Cash bank
Walk (bill)
Quick-change artist
Open check

Short (cash bank)
Over (cash bank)
Bonding (employee)
Collusion
General ledger
Cash receipts journal

Bank reconciliation
Receivable revenue
Auditor
Back office
  accounting system

Operator's 10-Point Tactics for Success Checklist

Evaluate your need for, and the current status of, each of the following operational tactics. For those tactics you think are important, but not yet in place, develop an action plan for its implementation including who will be responsible for the tactic's completion and the target date by which it should be completed.

				If Not Done	
Tactic	Don't Agree (Not Done)	Agree (Done)	Agree (Not Done)	Who Is Responsible?	Target Completion Date
1) Operator understands the importance to profits of developing an effective revenue control program.	——	——	——		
2) Operator can identify external threats to revenue security in their own operation.	——	——	——		
3) Operator can identify internal threats to revenue security in their own operation.	——	——	——		
4) Operator can state the objectives of an internal revenue security program.	——	——	——		
5) Operator can summarize the elements required in an effective revenue control system.	——	——	——		

Tactic	Don't Agree (Not Done)	Agree (Done)	Agree (Not Done)	If Not Done	
				Who Is Responsible?	Target Completion Date
6) Operator recognizes the importance of continually monitoring and verifying menu and bar item delivery to servers or guests.	——	——	——		
7) Operator recognizes the importance of continually monitoring and verifying guest charges.	——	——	——		
8) Operator recognizes the importance of continually monitoring and verifying sales receipts.	——	——	——		
9) Operator recognizes the importance of continually monitoring and verifying sales deposits.	——	——	——		
10) Operator recognizes the importance of continually monitoring and verifying the accounts payable (AP) system.	——	——	——		

12

Planning and Budgeting for Success

What You Will Learn

1) The Importance of Operating Budgets
2) How to Create an Operating Budget
3) How to Compare Actual Operating Results to Budgeted Results

Operator's Brief

In this chapter, you will learn the importance of creating and properly monitoring an operating budget. The operating budget, or financial plan, is developed to help a foodservice operation reach its future financial goals. Your operating budget tells you what you must achieve to meet your predetermined cost and profit objectives.

To create an operating budget, you must first consider:

1) Prior-period operating results (if yours is an existing operation)
2) Assumptions made about the next period's operation
3) The operation's financial objectives

An effective operating budget estimates your business's future sales, expenses, and profits for a specific accounting period. To estimate future sales most accurately, you must review historical records, and then consider any internal and external factors that may affect future revenue generation. The next step in the budgeting process is to estimate your fixed and variable costs.

You will learn that one important result of budget creation is the ability to make comparisons between budgeted results and your operation's actual financial performance.

In this chapter, you will learn how to analyze actual financial performance and compare it with the results you originally planned (budgeted). You will also learn how to use this information to take corrective action(s) or to modify your initial operating budget. Doing so will better ensure that you and your operation achieves the success you seek.

CHAPTER OUTLINE

The Importance of Operating Budgets
    Types of Operating Budgets
    Advantages of Operating Budgets
Creating an Operating Budget
    Revenue Forecasts
    Expense Forecasts
Monitoring the Operating Budget
    Comparing Planned Results to Actual Results
    Modifying the Operating Budget

The Importance of Operating Budgets

Just as the income statement (see Chapter 2) tells operators about their *past* performance, the **operating budget** (financial plan) helps a business predict and achieve its *future* financial goals. The operating budget tells foodservice operators what they must do to meet predetermined sales, costs, and profit objectives.

The utilization of an operating budget involves several activities:

Key Term

Operating budget: An estimate of the income and expenses of a business over a defined accounting period. Also referred to as a "financial plan."

1) Establishing realistic financial goals
2) Developing a budget (financial plan) to achieve financial goals
3) Comparing actual operating results with budgeted results
4) Taking corrective action, when necessary, to modify operating procedures and/ or the budget

Preparing a budget and staying within its financial boundaries help operators meet their financial goals. Without this plan, operators must guess the amounts to spend and how much sales should be anticipated. Effective operators build their budgets, monitor them closely, and modify them as necessary to achieve their desired results.

Types of Operating Budgets

One helpful way to consider the purpose of an operating budget is by its coverage (time frame). While an operating budget may be prepared for any accounting period desired, operating budget lengths are typically considered to be one of the three types shown in Figure 12.1.

Type of Operating Budget	Budget Characteristics
Long-range	Typically prepared for a period of up to five years. While not highly detailed, it provides financial views about long-term goals.
Annual	Typically prepared for one calendar or fiscal year or, in some cases, one season. The annual budget normally consists of 12 months or 13 periods of 28 days each.
Achievement	Prepared for a limited time, often a month, a week, or even one day. It most often provides very current operating information and greatly assists in making current operating decisions.

Figure 12.1 Operating Budgets

Advantages of Operating Budgets

The owners of a foodservice operation want to know what they should expect to earn from their investments, and a budget helps to project these earnings. Questions about the amount of revenue likely generated, the amount of cash that should be available for bill payment or distribution as earnings, and the proper timing of major purchases can all be addressed in a properly developed budget.

The advantages of preparing and using an operating budget are many and are summarized in Figure 12.2.

1) It is a proven plan to analyze alternative courses of action, and it allows operators to examine alternatives before adopting a specific plan.
2) It requires operators to examine the facts about what their desired profit levels should be.
3) It enables operators to define standards used to develop and enforce appropriate cost control systems.
4) It allows operators to anticipate and prepare for future business conditions.
5) It helps operators periodically carry out a self-evaluation of their business and its progress toward meeting financial objectives.
6) It is a communication channel that allows the operation's objectives to be passed along to stakeholders including owners, investors, managers, and staff.
7) It encourages those who participated in budget preparation to establish their own operating objectives, evaluation tactics, and tools to implement operating budgets.
8) It provides operators with reasonable estimates of future expense levels and serves as an important aid when appropriate selling prices are determined.
9) It identifies time periods in which operational cash flows may need to be supplemented.
10) It communicates realistic financial performance expectations of a business's operators to the business's owners and investors.

Figure 12.2 Advantages of Preparing and Using an Operating Budget

One good way to consider operating budgets is to compare them to an operation's income statement. Recall that the income statement details the actual revenue, expenses, and profits incurred in operating a business. The operating budget is actually the operator's best estimate of all or any portion of a future income statement.

Find Out More

Some foodservice operators responsible for budget preparation for one or more units find membership in the Hospitality Financial and Technology Professionals (HFTP®) to be helpful.

Established in 1952, HFTP is an international, nonprofit association headquartered in Austin, Texas, with offices in the United Kingdom, Netherlands, and Dubai. HFTP is recognized as the professional group representing the finance and technology segments of the hospitality industry. It offers members continuing education courses including those addressing budgeting through its HFTP Academy.

To find out more about the educational resources offered by this professional group and its "Certified Hospitality Accountant Professional (CHAE)" program, enter "HFTP Academy" in your favorite search engine and view the results.

Creating an Operating Budget

Some operators believe it is difficult to develop an operating budget, and they do not take the time to do so. Creating an operating budget, however, does not need to be a complex (i.e., time consuming) process.

Before operators can begin to develop a budget, they must understand the essentials required for its creation. If these are not addressed before the operating budget is developed, the budgeting process that follows is not likely to yield an accurate or helpful financial planning tool.

Before beginning the creation of an operating budget, developers must have available and understand the following:

1) Prior-period operating results (if an existing operation)
2) Assumptions about the next period's operations
3) Knowledge of the organization's financial objectives

1) Prior-Period Operating Results

The task of budgeting is easier when an operator knows the results of prior accounting periods. Experienced foodservice operators know that what occurred in their units in the past is often a good indicator of what may occur in the future. The further back and the greater detail with which an operator can track historical revenues and expenses, the more accurate budget planning can be.

For example, if operators know the revenues and expenses for the past 50 Saturdays, they are better able to forecast this coming Saturday's revenue and expense budgets than if only the last two Saturdays' data is available.

When preparing a budget, **historical data** should always be considered along with the most recent data available.

Key Term

Historical data: Information about an operation's past financial performance including revenue generated, expenses incurred, and profits (or losses) realized.

For example, assume a foodservice operator knows revenues have, on average, increased 5% each month from the same period last year. However, in the last two months, the increase has been closer to zero. This may mean that the revenue increase trend has slowed or even stopped completely. Successful operators modify historical trends by closely examining current conditions. In this example, the operator should probably estimate next month's revenue increases to be closer to 0% than to 5%. Note: While historical data is useful for ongoing operations, this data will not be available for new operations.

2) Assumptions about the Next Period's Operations

Evaluating future conditions and business activity is always a key part of operating budget development. Examples include the opening of new competitive restaurants in the immediate area, scheduled occurrences including local sporting events, festivals and concerts, and significant changes in operating hours. Local newspapers, trade or business associations, and Chambers of Commerce are possible sources of helpful information about changes in future demand for an operation's products and services.

When significant changes are planned for an operation such as the introduction of new menu items, changes in operating hours, or the estimated impact of significant marketing efforts, assumptions about the impact of these actions become important. After these factors have been considered, realistic assumptions about changes in potential revenues and expenses may be made.

3) Knowledge of the Organization's Financial Objectives

An operation's financial objectives may consist of a specific profit target such as a percentage of revenue or a total dollar amount. Some financial objectives are determined by an operation's owner based on desired ROI (see Chapter 2) and the operating budget must also address these goals.

The operating budget is actually a detailed plan that can be expressed by the basic formula:

Budgeted Revenue − Budgeted Expense = Budgeted Profit

The budgeted profit level an operator desires can be achieved when the operation realizes its budgeted revenue levels and spends only what is budgeted to generate the sales. If revenues fall short of forecast and/or if expenses are not reduced to match the shortfall, budgeted profit levels are not likely to be achieved.

In a similar manner, if actual revenues exceed forecasted levels, some expenses will also increase. If the increases are monitored carefully and are not excessive, increased profits should result. If, however, an operator allows actual expenses to exceed the levels required by the additional revenue, budgeted profits may not be achieved.

To illustrate the operating budget development process, consider Gene, the owner/operator of Gene's Restaurant. He is developing the operating budget for next year and has determined historical budget data essentials as shown below:

1) He has gathered his prior year operating results.
2) From information applicable to the area's economic conditions and his competitors, he has made the following assumptions about next year's operations:
   - Total revenues received will increase by 4% primarily because of a 4% menu price increase.
   - Food and beverage costs will increase by 3% due to inflation affecting food and beverage product prices. As a result, his product expense (cost of sales) targets are a 35% food cost and a 16% beverage cost. This will generate a combined weighted food and beverage total cost of sales of 31.2% (food cost/food sales + beverage cost/beverage sales = 31.2%).
   - Management, staff wages, and benefits costs have a target of 35% of sales for total labor cost.
   - Prime cost will be 66.2% of sales.
   - All other controllable expenses will total no more than 12.4% of sales.
3) Gene's financial objectives for the restaurant are to earn profits (net income before income taxes) of at least 11.1% of sales, and net income (after taxes) of 8.3% of sales.

Given these assumptions, Gene is now ready to create each of the 12-monthly revenue and expense forecasts needed to complete next year's operating budget.

Revenue Forecasts

Accurately forecasting an operation's revenues is critical because all forecasted expenses and profits are based on revenue forecasts. In most cases, revenues should be estimated on a monthly (or weekly) basis. These estimates can then be combined to create the annual revenue budget. The reason: Many hospitality operations have seasonal revenue variations.

For example, a restaurant doing a significant amount of business in its outside dining patio area may generate reduced sales during times of the year when inclement weather makes outside dining less desirable for guests. Similarly, a restaurant operated in a ski resort town will generally be busier in the winter ski season than in summer months.

Forecasting revenues is not an exact science. However, it can be made more accurate when operators:

✓ **Review historical records.** Operators begin the revenue forecasting process by reviewing their revenue records from previous years. When an operation has been open at least as long as the budget period being developed, its revenue history can be extremely helpful in predicting future revenue levels.

✓ **Consider internal factors affecting revenues.** In this step, operators consider any significant changes in the type, quantity, and direction of their marketing efforts. Other internal activities that can impact future revenues include those related to facility renovation affecting dining capacity, the number of hours to be open, and/or changes in menu prices. Any internal change an operator believes will likely impact future revenues should be considered in this step.

✓ **Consider external factors affecting revenues.** There are numerous external issues that could affect an operation's revenue forecasts. These include planned competitors' openings or closings, and other factors such as road improvements or construction that disrupts normal traffic patterns. Other factors that may yield revenue forecast changes can relate to forecasted economic upturns or downturns that affect how potential guests may spend their money on dining services.

Returning to the example of Gene's Restaurant and using September as an example of the month he is now working on, Gene has reviewed last year's data for the month of September and found that his sales that month were $192,308, with a guest check average (see Chapter 2) of $12.02.

He considers the internal and external factors affecting revenues and has estimated a net 4% increase in revenues for the coming September. Gene then computes his revenue forecast for the upcoming September as:

Sales Last Year × (1.00 + % Increase Estimate) = Revenue Forecast

Or

$192,308 × (1.00 + 0.04) = $200,000$

Using his historical data, Gene knows that approximately 80% of his sales are from food, and 20% of his sales are from beverages. Therefore, he estimates $160,000 ($200,000 × 0.80 = $160,000) for food sales and $40,000 ($200,000 × 0.20 = $40,000) for beverage sales.

Gene's average sale per guest (see Chapter 2) for food and beverages for last September was $12.02. With a forecasted increase of 4% in selling prices, the forecast for this September's average sale per guest (check average) is calculated as:

Guest Check Average Last Year × (1 + % Increase Estimate)
= Guest Check Average Forecast

Or

$12.02 × (1.00 + 0.04) = 12.50

With forecasted sales of $200,000 and a forecasted guest check average of $12.50, Gene's forecasted number of guests would be 16,000 ($200,000 revenue ÷ $12.50 guest check average = 16,000 guests).

In most cases, it is not realistic to assume an operator can forecast their business's exact monthly revenue one year in advance. With accurate historical sales data, and a realistic view of internal and external variables that can impact an operation's future revenue generation, many operators can attain operating budget forecasts that are within 5–10% (plus or minus) of their actual results for a forecasted accounting period.

Technology at Work

Today's readily available restaurant sales forecasting tools help foodservice operators and owners better estimate future volume and make more informed and accurate decisions about purchasing and staffing.

These advanced technology tools, often interfaced with an operation's point-of-sale (POS) system, can use historical data to help foresee upcoming seasonality trends and predict how the trends will likely affect an operation's revenue generation. Since they assist in establishing realistic sales and profit goals, the result is better financial planning and preparation for the future.

To examine information about some systems that selected companies offer to assist in foodservice sales forecasting, enter "restaurant revenue forecasting software" in your favorite search engine and review the results.

Expense Forecasts

Operators must budget for each fixed and variable cost when they address individual expense categories on income statements. Fixed costs are simple to forecast because items such as rent, depreciation, and interest typically do not vary from month to month.

Variable costs, however, are directly related to the amount of revenue produced by a foodservice operation. For example, an operation that forecasts sales of 100 prime rib dinners on Friday night will have higher food and server (labor) costs than the operator in a similar facility that forecasts the sale of only 50 prime rib dinners. The reason: Variable expenses such as food, beverage, and labor costs are affected by sales levels.

Based on the information he has gathered, the resulting budget for Gene's Restaurant for September of the budget year is shown in Figure 12.3.

Gene's Restaurant

Budget for September Next Year

Budgeted Number of Guests = 16,000

	Next Year	%
SALES		
Food	$ 160,000	80.0
Beverage	$ 40,000	20.0
Total Sales	$ 200,000	100.0
COST OF SALES		
Food	$ 56,000	35.0
Beverages	$ 6,400	16.0
Total Cost of Sales	$ 62,400	31.2
LABOR		
Management	$ 18,000	9.0
Staff	$ 38,000	19.0
Employee Benefits	$ 14,000	7.0
Total Labor	$ 70,000	35.0
PRIME COST	$ 132,400	66.2
OTHER CONTROLLABLE EXPENSES		
Direct Operating Expenses	$ 7,856	3.9
Music & Entertainment	$ 1,070	0.5
Marketing	$ 3,212	1.6
Utilities	$ 5,277	2.6
General & Administrative Expenses	$ 5,570	2.8
Repairs & Maintenance	$ 1,810	0.9
Total Other Controllable Expenses	$ 24,795	12.4
CONTROLLABLE INCOME	$ 42,805	21.4
NONCONTROLLABLE EXPENSES		
Occupancy Costs	$ 10,000	5.0
Equipment Leases	$ —	0.0
Depreciation & Amortization	$ 3,400	1.7
Total Noncontrollable Expenses	$ 13,400	6.7
RESTAURANT OPERATING INCOME	$ 29,405	14.7
Interest Expense	$ 7,200	3.6
INCOME BEFORE INCOME TAXES	$ 22,205	11.1
Income Taxes	$ 5,551	2.8
NET INCOME	$ 16,654	8.3

Figure 12.3 Gene's Restaurant Operating Budget for September Next Year

As shown in Figure 12.3, every expense, considered individually, must be included in the operating budget. When properly completed, the result will be an operating budget that:

1) Is based upon a realistic revenue estimate
2) Considers all known fixed and variable costs
3) Is intended to achieve the organization's financial goals
4) Can be monitored to ensure adherence to the budget's guidelines
5) May be modified, when necessary

Annual operating budgets are most often the compilation of monthly operating budgets. Therefore, operators may use their monthly operating budgets to create annual budgets and monitor weekly (or even daily) versions of their overall operating budgets.

What Would You Do? 12.1

"I just don't know," said Trishauna, "it might make an impact, but it might not make a very big one. I think we'll just have to wait and see."

Trishauna was talking to Nora, her partner in the Jacked Up Coffee Bar, a coffee and pastry shop located adjacent to the State University campus.

Trishauna and Nora were preparing next year's operating budget. They had just begun the process and were forecasting their next year's revenues. Both Trishauna and Nora were aware that a major coffee shop chain had just announced plans to open a new shop within one block of the Jacked Up Coffee Bar.

"Well," said Nora "I think our clientele is loyal. I know our new competitor will generate some business, but I'm not sure how much of that business will come from us versus other coffee shops in the area."

"Exactly," replied Trishauna. "That's why I think it'll be a real challenge for us to forecast our revenue for next year."

Assume you were the owners of the Jacked Up Coffee Bar. How important will it be for you to consider external influences such as the opening of a new competitor as you create your next year's revenue budget? What could be a likely result if you did not make such considerations?

Monitoring the Operating Budget

An operating budget details a plan for future financial activities. In many cases, an operator's forecast of future results will be reasonably accurate but, in other cases, it will not be as accurate. For example, revenue may not reach forecasted levels, expenses may exceed estimates, and internal or external factors not considered when the operating budget was prepared may negatively or positively impact financial performance.

An operation's budget will have little value if management does not utilize it. The operating budget should be regularly monitored in each key area:

✓ Revenue
✓ Expenses
✓ Profit

Revenue

If sales should fall below projected levels, the impact on profits can be substantial, and it may not be possible to meet profit goals. If revenues consistently exceed projections, variable cost portions of the budget must be modified or, ultimately, these expenses will soon exceed their budgeted dollar amounts. Effective operators compare their actual sales to those they have projected on a regular basis.

An appropriate revenue comparison must include a comparison of total sales and guest check averages. The reason: An operation's sales may increase at the same time its guest count is increasing, or an operation's sales may increase with a declining guest count (if the average sale per guest increases significantly). While both situations will result in an increase in total revenue, it is important for foodservice operators to understand the sources of their revenue increases.

Expenses

Foodservice operators must be careful to monitor their operating expenses because costs that are too high or too low may be cause for concern. Just as it is not possible to estimate future sales volumes perfectly, it is also not possible to estimate future expenses perfectly, especially since some expenses will vary as sales volumes increase or decrease.

As business conditions change, revisions in the operating budget are to be expected. This is true because operating budgets are based on a specific set of assumptions and, as the assumptions change, so will the accuracy of the operating budget produced from the assumptions.

To illustrate, assume an operator budgeted $1,000 in January for snow removal from the parking lot attached to a restaurant operating in upper New York State. If unusually severe weather causes the operator to spend $2,000 for snow removal in January, the assumption (normal levels of snowfall) was incorrect, and the original budget will be incorrect as well.

Profit

An operation's forecasted profits must be realized if the operation is to successfully provide adequate returns for its owner(s). To illustrate, consider Latisha, the manager of a foodservice establishment with excellent sales but below-budgeted profits. For this year, Latisha budgeted a 5% profit on $2,000,000 of sales and, therefore, a $100,000 profit ($2,000,000 × 0.05 = $100,000) was anticipated.

At year's end, Latisha achieved her sales goal, but in doing so she generated only $50,000 profit, or 2.5% of sales (($50,000 ÷ $2,000,000) × 100 = 2.5%)). If this operation's owners feel that $50,000 is an adequate return for their investment and risk, Latisha's services may be retained. If they do not, she may lose her position, even though she operates a "profitable" restaurant. The reason: Management's task was not merely to generate a profit, but rather to generate the realistic profit level that was planned.

Comparing Planned Results to Actual Results

One important task of a foodservice operator is to optimize profitability by analyzing the differences between planned for (budgeted) results and actual operating results. To do this effectively, operators must receive timely income statements that accurately detail what occurred (their actual operating results). When they do, the actual results can then be compared to the operating budget for the same accounting period. Recall from Chapter 3 that a variance is the difference between planned (budgeted) financial results and actual financial results.

The basic formula operators use to calculate a budget variance is:

Actual Results − Budgeted Results = Variance

A variance may be expressed in either dollar or percentage terms, and it can be either positive (favorable) or negative (unfavorable). A **favorable variance** occurs when the variance is an improvement on the budget (revenues are higher or expenses are lower). An **unfavorable variance** occurs when actual results do not meet budget expectations (revenues are lower or expenses are higher).

For example, if the budget for carpet cleaning is $1,000 for a given month, but the actual expenditure for those services is $1,250, the variance is calculated as:

Actual Results − Budgeted Results = Variance

Or

$1,250 − $1,000 = $250

Key Term

Favorable variance: Better-than-expected performance when actual results are compared to budgeted results.

Key Term

Unfavorable variance: Worse-than-expected performance when actual results are compared to budgeted results.

In this example, the variance may be expressed as a dollar amount ($250) or as a percentage of the original budget. The computation for a percentage variance is:

$$\frac{\text{Variance}}{\text{Budgeted Results}} \times 100 = \text{Percentage Variance}$$

Or

$$\frac{\$250}{\$1,000} \times 100 = 25\%$$

In this example, the variance is unfavorable to the operation because the actual expense is higher than the budgeted expense. In business, a variance can also be considered as either significant or insignificant. It is the operator's task to identify and address significant variances between budgeted and actual operating results.

A **significant variance** may be defined in several ways. A common definition is that a significant variance is any difference in dollars or percentage between budgeted and actual operating results that warrants further investigation. Significant variance is an important concept because not all variances must be investigated.

Key Term

Significant variance: A variance that requires immediate management attention.

For example, assume that, at the beginning of a year, a foodservice operator prepares an annual (12-month) operating budget that forecasts the operation's December utility bill will be $6,000. When, 12 months later, the December bill arrives, it totals $6,420 and, therefore it is $420 over budget ($6,420 actual expense − $6,000 budgeted expense = $420 variance).

Given the amount of the bill ($6,420) and the difficulty of accurately estimating utility expenses one year in advance, a difference of only $420 or 7% (($420 ÷ $6,000) × 100 = 7.0%)) probably does *not* represent a significant variance from the operation's budget.

Alternatively, assume that the same operation had estimated office supplies usage at $100 for that same month, but the actual cost of supplies was $520. Again, the difference between the budgeted expense and actual expense is $420 ($520 actual − $100 budgeted = $420 variance). The office supplies variance, however, represents a very significant difference of 420% (($420 ÷ $100) × 100 = 420%)) between planned and actual results. This situation should probably be thoroughly investigated.

Foodservice operators must determine what represents a significant variance based on knowledge of their specific operation. Small percentage differences can be important if they represent large dollar amounts. Similarly, small dollar amounts can be significant if they represent large percentage differences from planned results. Variations from budgeted results can occur in revenues, expenses, and profits. Operators can monitor all these areas using a four-step operating budget monitoring process.

Step 1: Compare actual results to the operating budget
Step 2: Identify significant budget variances
Step 3: Determine causes of significant budget variances
Step 4: Take corrective action or modify the operating budget

In Step 1, an operator reviews their income statement and operating budget data for a specified accounting period. In Steps 2 and 3, actual operating results are compared to the budget and significant variances, if any, are identified and analyzed. Finally, in Step 4, corrective action is taken to reduce or eliminate unfavorable variances or, if it is appropriate to do so, the budget is modified to reflect the new realities confronting the business.

In most cases, operators should compare their actual results to their operating budget results in each of the income statement's three major sections (sales, expenses, and profits). To illustrate the process, consider again the operating budget for Gene's Restaurant. Figure 12.4 shows Gene's original operating budget (see Figure 12.3) and his actual operating results in dollars and percentages of sales for the month of September.

If Gene properly monitors his budget, sales (revenue) is the first area he will examine when comparing his actual results to his budgeted results. The reason: If sales fall significantly below projected levels, there will likely be a significant negative impact on profit goals.

Secondly, when sales vary from projections, variable costs will also fluctuate. In cases where sales are lower than budget projections, variable costs should most often be *less* than budgeted. In addition, when actual sales fall short of budgeted levels, fixed expenses such as rent, and labor incurred by the operation will represent a larger-than-originally budgeted *percentage* of total sales. Alternatively, when actual sales exceed the budget, the total dollar value of variable expenses will increase. However, fixed expenses should, if properly managed, represent a smaller-than-budgeted percentage of total sales.

A close examination of Figure 12.4 shows that Gene has experienced a shortfall in both food and beverage revenue when compared to his operating budget. One revenue-related problem Gene faced in September is that he budgeted for 16,000 guests and only served 15,500 guests.

Also, Gene budgeted for a $12.50 average guest check ($200,000 budgeted sales/16,000 budgeted guests = $12.50 budgeted guest check average). However,

Gene's Restaurant

Budget versus Actual Comparison for September

	Budgeted Number of Guests = 16,000		Actual Number of Guests = 15,500	
	Budget	**%**	**Actual**	**%**
SALES				
Food	$160,000	80.0%	$150,750	79.3%
Beverage	$ 40,000	20.0%	$ 39,250	20.7%
Total Sales	$200,000	100.0%	$190,000	100%
COST OF SALES				
Food	$ 56,000	35.0%	$ 53,800	35.7%
Beverages	$ 6,400	16.0%	$ 6,300	16.1%
Total Cost of Sales	$ 62,400	31.2%	$ 60,100	31.6%
LABOR				
Management	$ 18,000	9.0%	$ 18,000	9.5%
Staff	$ 38,000	19.0%	$ 37,800	19.9%
Employee Benefits	$ 14,000	7.0%	$ 11,160	5.9%
Total Labor	$ 70,000	35.0%	$ 66,960	35.2%
PRIME COST	$132,400	66.2%	$127,060	66.9%
OTHER CONTROLLABLE EXPENSES				
Direct Operating Expenses	$ 7,856	3.9%	$ 7,750	4.1%
Music & Entertainment	$ 1,070	0.5%	$ 1,070	0.6%
Marketing	$ 3,212	1.6%	$ 1,350	0.7%
Utilities	$ 5,277	2.6%	$ 5,195	2.7%
General & Administrative Expenses	$ 5,570	2.8%	$ 5,455	2.9%
Repairs & Maintenance	$ 1,810	0.9%	$ 1,925	1.0%
Total Other Controllable Expenses	$ 24,795	12.4%	$ 22,745	12.0%
CONTROLLABLE INCOME	$ 42,805	21.4%	$ 40,195	21.2%
NONCONTROLLABLE EXPENSES				
Occupancy Costs	$ 10,000	5.0%	$ 10,000	5.3%
Equipment Leases	$ —	0.0%	$ —	0.0%
Depreciation & Amortization	$ 3,400	1.7%	$ 3,400	1.8%
Total Noncontrollable Expenses	$ 13,400	6.7%	$ 13,400	7.1%
RESTAURANT OPERATING INCOME	$ 29,405	14.7%	$ 26,795	14.1%
Interest Expense	$ 7,200	3.6%	$ 7,200	3.8%
INCOME BEFORE INCOME TAXES	$ 22,205	11.1%	$ 19,595	10.3%
Income Taxes	$ 5,551	2.8%	$ 4,899	2.6%
NET INCOME	$ 16,654	8.3%	$ 1 4,696	7.7%

Figure 12.4 Gene's Restaurant Budget versus Actual Comparison for September

he achieved a guest check average of only $12.26 ($190,000 actual sales/15,500 actual guests = $12.26 actual guest check average).

If sales consistently fall short of forecasts, Gene must evaluate all aspects of his entire operation to identify and correct the revenue shortfalls. Foodservice operations that consistently fall short of their sales projections must be evaluated to learn the validity of the primary assumptions used to produce the sales portion of their operating budgets.

Find Out More

To achieve their desired sales levels, foodservice operators must have an effective marketing plan in place. Increasingly, an operation's marketing efforts must include both traditional approaches and newer Internet-based approaches.

While there are many publications addressing general marketing strategies for businesses, few publications exclusively address the marketing needs of foodservice operations.

One of the best and most up-to-date marketing resources available to foodservice operators, and one that exclusively addresses the on-site and online marketing of foodservice operations is Marketing in Foodservice Operations, part of the "Foodservice Essentials" series published by John Wiley.

To learn more about the content and availability of this extremely valuable publication, go to www.wiley.com. When you arrive at the Wiley website, enter "Marketing in Foodservice Operations" in the search bar and review the results.

Expense Analysis

Identifying significant expense variances is perhaps the most critical part of the budget monitoring process because many types of operating expenses are controllable. Some variation between budgeted and actual costs can be expected because most variable operating expenses vary with sales levels that cannot be predicted perfectly. The variances that occur can, however, tell operators a great deal about operational efficiencies, and experienced operators know that a key to ensuring profitability is to properly examine and manage controllable costs.

As shown in Figure 12.4, Gene's cost of food in dollars were lower than budgeted ($56,000 budgeted vs. $53,800 actual) as he would expect given his food sales shortfall. However, his cost of sales percentage was over budget (35.0% budgeted vs. 35.7% actual), and this is an indication that his kitchen was not as cost-efficient as he would have liked. The same was true in his beverage sales category. As a result, a close examination of his product cost percentages will be an important part of Gene's analysis of budgeted versus actual expenses.

As was true in this example of Gene's Restaurant, operators may find that their product costs, when expressed as a percentage of sales, are too high, and they must be reduced. Operators facing this situation can choose from a variety of solutions to this problem if they first carefully consider the cost of sales equation.

Recall from Chapter 7 that the formula operators use to calculate their food cost percentage is:

$$\frac{\text{Cost of Food Sold}}{\text{Food Sales}} = \text{Food Cost \%}$$

Similarly, the formula used for calculating a beverage cost percentage is:

$$\frac{\text{Cost of Beverage Sold}}{\text{Beverage Sales}} = \text{Beverage Cost \%}$$

In its simplest terms, both of these product cost percentages can be expressed algebraically as:

$$\frac{A}{B} = C$$

where:

A = Cost of Products Sold
B = Sales
C = Cost Percentage

Figure 12.5 shows how algebra affects the equation and what the rules of algebra communicate to foodservice operators.

In general, foodservice operators must control the variables that impact the product cost percentage and reduce the overall value of "C" in the product cost percentage equation. Basic product cost reduction approaches can be used to optimize the overall product cost percentage. These approaches are to:

✓ Ensure all purchased products are sold
✓ Decrease portion size relative to selling price
✓ Vary recipe composition
✓ Alter product quality
✓ Achieve a more favorable sales mix
✓ Increase selling price relative to portion size

These strategies can be applied when excessive food and/or beverage costs exist. It is the careful selection and mixing of these approaches to cost control that differentiates successful operators from their less successful counterparts.

Algebraic Rule	Foodservice Operator Takeaway
If A is unchanged and B increases, C decreases.	If product costs can be kept constant while sales increase, the product cost percentage goes down.
If A is unchanged and B decreases, C increases.	If product costs remain constant but sales decline, the cost percentage increases.
If A increases at the same proportional rate B increases, C remains unchanged.	If product costs go up at the same rate that sales go up, the cost percentage will remain unchanged.
If A decreases and B is unchanged, C decreases.	If product costs can be reduced while sales remain constant, the cost percentage goes down.
If A increases and B is unchanged, C increases.	If product costs increase with no increase in sales, the cost percentage will go up.

Figure 12.5 Rules of Algebra and Foodservice Operator Takeaways

It is not the authors' contention that there are no other possible product cost reduction methods. However, the alternative approaches presented here illustrate how a careful analysis of cost percentage reduction alternatives can benefit all operators.

✓ **Ensure All Purchased Products Are Sold**

This strategy has tremendous implications, and it includes all phases of professional purchasing, receiving, storage, inventory management, issuing, production, service, and cash control. Perhaps the hospitality industry's greatest challenge in product cost control is ensuring that all products, once purchased, generate cash sales that are ultimately deposited into an operation's bank account.

✓ **Decrease Portion Size Relative to Selling Price**

Product cost percentages are directly affected by an item's portion cost (see Chapter 7) which is a direct result of portion size. Too often, foodservice and bar operators assume their standard portion sizes must conform to some unwritten rule of uniformity. However, most guests would prefer a smaller portion size of higher quality ingredients than the reverse. In fact, one problem some restaurants must address are that their portion sizes are too large. The result: excessive food loss because uneaten products left on plates must be thrown away. Remember that foodservice operators solely determine portion sizes, and they are variable.

An example of the impact of portion size on the product cost percentage is shown in Figure 12.6. This figure presents the significant effect on the liquor cost percentage of varying the standard drink size assuming:

1) $21.00 per liter as the standard cost of the liquor
2) A 0.8-ounce evaporation rate per liter, resulting in 33 ounces of salable liquor per liter
3) A standard $10.00 selling price per drink

Note in this example that the operator's liquor cost percentage ranges from 6.4% of sales when a 1-ounce portion is served to a 12.7% product cost when a 2-ounce portion is served.

Portion sizes of both food and drink items directly affect product cost percentages and, as well, the guests' perceptions of value delivered. As a result, when establishing portion sizes, operators must carefully consider all variables affecting their sales. These may include location, service levels, competition, and the type of clientele being served.

✓ Vary Recipe Composition

Experienced foodservice operators know that even the simplest standardized recipes can often be changed. For example, consider the proper ratio of clams to potatoes when preparing 100 servings of high-quality clam chowder. Since the cost of one pound of clams far exceeds the cost of one pound of potatoes, 100 servings of clam chowder made with an increased amount of clams will cost more to produce than 100 servings of chowder made with increased amounts of potato. What constitutes an ideal recipe composition in this example? The answer must be addressed by the operator, and the "answer" in each standardized recipe used will impact the operation's overall food cost percentage.

Similarly, the proportion of alcohol to mixer affects beverage cost percentages. Sometimes the amount of alcohol in drinks can be reduced, and overall

Drink Size	Drinks per Liter	Cost Per Liter	Cost Per Drink	Sales Per Liter	Liquor Cost % Per Liter
2 oz.	16.5	$21.00	$1.27	$165.00	12.7%
1¾ oz.	18.9	21.00	$1.11	$189.00	11.1%
1½ oz.	22.0	21.00	$0.95	$220.00	9.5%
1¼ oz.	26.4	21.00	$0.80	$264.00	8.0%
1 oz.	33.0	21.00	$0.64	$330.00	6.4%

Figure 12.6 Impact of Drink Size on Liquor Cost Percentage at Constant Selling Price of $10.00 per Drink

drink sizes can be increased. One example: Increasing the drink's proportion of lower cost standardized drink recipe ingredients such as milk, juices, and soda may enable the use of a smaller portion of higher cost spirit products. Utilizing this beverage cost reduction strategy might contribute to a feeling of satisfaction by the guest while, at the same time, allowing the operator to reduce beverage cost percentages and increase profitability.

✓ **Alter Product Quality**

In nearly all cases, higher quality food and beverage products cost more than lower quality products. Therefore, one way to reduce product costs is to reduce product quality. This area must be approached with great caution because a foodservice operation should never serve products of unacceptable quality. Rather, cost conscious operators should always purchase the quality of product appropriate for its intended use.

When operators determine that an appropriate ingredient, rather than the highest cost ingredient, provides good quality and value to guests, product costs might be reduced with product substitution. One caveat: Lower quality products may cost less, but customers may perceive that menu items made from these lower quality ingredients result in reduced levels of value.

✓ **Achieve a More Favorable Sales Mix**

Experienced operators know that their customers' menu item selection decisions have a direct and significant impact on product cost percentages. The reason: An operation's overall product cost percentage is determined in large part by the operation's **sales mix**.

Key Term

Sales mix: The series of individual guest purchasing decisions that result in a specific overall food or beverage cost percentage.

Sales mix affects the overall product cost percentage anytime guests have a choice among several menu selections, and each selection has its own unique product cost percentage.

To consider how the sales mix directly affects an operation's overall product costs, assume that only three different menu items are sold in this foodservice operation. Each item is priced separately, but the operation also uses bundling (see Chapter 7) to produce a "Bundle Meal." Note: This meal includes one of each item when purchased at the same time as shown in Figure 12.7.

When reviewing Figure 12.7, it is easy to see that if, on a specific day, 100% of the operation's guests bought a hamburger and nothing else, the operation's overall product cost would be 37.6%. If, on another day, 100% of its customers purchased a soft drink and nothing else, the product cost percentage for that day would be 15.2%.

Menu Item	Product Cost	Selling Price	Product Cost %
Hamburger	$1.50	$3.99	37.6%
French fries	$0.50	$1.99	25.1%
Soft drink	$0.15	$0.99	15.2%
Bundle Meal	$2.15	$6.49	33.1%

Figure 12.7 Four-item Menu

Similarly, if every guest purchased only the "Bundle Meal" on a specific day, the operation would achieve a 33.1% product cost. An operation's actual product cost is largely determined by the "mix" of the individual product costs resulting from the menu item choices made by the operation's guests.

Operators can directly influence guest selection and sales mix by techniques including strategic pricing, effective menu design, and creative marketing. However, guests will always determine an operation's overall product cost percentage because their orders produce an operation's unique sales mix.

Recognizing the impact of sales mix can help operators better understand that effective sales mix management can reduce product cost percentages. It can also increase profitability while allowing portion size, recipe composition, and product quality to remain constant.

✓ Increase Selling Price Relative to Portion Size

Some operators that are confronted with rising product costs and increased cost percentages think the only appropriate response is to increase their menus' selling prices. While this can be done, the tactic must be approached with caution. There may be no larger foodservice temptation than to raise prices to counteract management's ineffectiveness at controlling product costs!

There are times when selling prices must be increased, and this is especially true when there is inflation and unique product shortages. Price increases should be considered, however, only when all other alternatives and necessary steps to control product costs have been considered and effectively implemented. For example, assume an operation has a fresh orange slice as part of a garnish on a popular fresh salad. If the cost of oranges increases significantly, perhaps the garnish can be made with another fruit during the time oranges are expensive.

Returning to the example of Gene's Restaurant, a review of Figure 12.4 reveals that Gene's total dollar amount spent for labor (management, staff, and employee benefits) was lower than budgeted ($70,000 budgeted vs. $66,960 actual). However, his reduced sales level resulted in a slight increase in total labor percentage (35.0% budgeted vs. 35.2% actual).

Note also in Figure 12.4 that Gene's fixed costs (e.g. occupancy, depreciation, and interest) did not vary in dollar amount. However, because revenues did not reach their budgeted levels, these also represented a higher actual cost percentage of sales because the fixed dollars were spread over a smaller revenue base.

The marketing expense category in Figure 12.4 is an expense line item that illustrates the need to carefully compare budgeted expense to actual expense. Gene's September marketing budget was $3,212. His actual marketing expense was $1,350. This variance might initially seem to be positive because less was spent on this expense than was previously budgeted. However, it is possible that that some of Gene's shortfall in revenue may have been created because he reduced his marketing expense. Experienced foodservice operators know that savings achieved in marketing costs often result in no savings at all!

Note also that Controllable Income in Figure 12.4 was budgeted at 21.4% of total revenue, but the actual results were lower (21.2%). The difference in the dollar amount of Controllable Income achieved was $2,610 ($42,805 budgeted − $40,195 actual = $2,610 variance). Gene must decide if this constitutes a "significant" variance. If so, it should be analyzed using the four-step "Operating Budget Monitoring Process" presented earlier in this chapter.

Profit (Net Income) Analysis

A foodservice operation's actual level of profit is measured either in dollars, percentages (or both), and it is the most critical number that most operators evaluate. Returning to Figure 12.4, it is easy to see that Gene's actual net income (the "bottom line") for the month was $14,696 or 7.7% of total sales. This is less than the $16,654 (8.3% of total sales) that was forecast in the operating budget. This reduced profit level can be tied directly to Gene's lower-than-expected sales.

When a business is unable to meet it topline revenue (see Chapter 2) forecasts, it typically means the budget was ineffectively developed, internal/external conditions have changed, and/or the operation's marketing efforts were not effective. Regardless of the cause, when sales do not reach forecasted levels, corrective action may be needed to prevent even more serious problems including significant future profit erosion.

Foodservice profits (net income) are routinely reported as both dollars and percentages, but foodservice operators may disagree about the best way to evaluate their profitability. One frequently hears comments in the hospitality industry such as "You bank dollars, not percentages!" To better understand this statement, note in Figure 12.4 that Gene's net income was a lower percentage than the operating budget forecast (7.7% actual net income vs. 8.3% budgeted net income).

However, consider the hypothetical results that would be obtained if Gene's actual net income percentage in September had been 8.5%. This would have been a *higher* percentage than his operating budget predicted (8.5% actual vs. 8.3% budgeted).

In this hypothetical scenario, the $16,150 total dollars of profit that would be generated ($190,000 actual sales × 8.5% net income = $16,150) would still be *less* than the initially budgeted 8.3% profit of $16,654. This, then, is the reasoning behind the statement, "You bank dollars, not percentages!"

Modifying the Operating Budget

Successful operators know their operating budgets should be an active and potentially evolving document. The budget should be regularly reviewed and, as necessary, modified as new and better information replaces that available when the original operating budget was developed. This is especially true when new data significantly (and perhaps permanently) affects the sales and expense assumptions used to create the budget. The operating budget should be reviewed anytime an operator believes the assumptions upon which it was based are no longer valid.

To illustrate, assume a foodservice operator employs 25 full-time employees. Each employee is covered under the operation's group health insurance policy. Last year, the operator agreed to pay 50% of each employee's insurance cost and, as a result, paid $300 per month for every full-time employee. The total cost of the insurance contribution each month was $7,500 (25 employees × $300 per employee = $7,500).

When this year's budget was developed, the operator assumed a 10% increase in health insurance premiums. If, later in the year, it is determined that premiums will be increased by 20%, employee benefit costs will be much greater than that projected in the original operating budget. This operator now faces several choices:

✓ Modify the budget.
✓ Reduce the amount contributed per employee to stay within the budget.
✓ Change (reduce) health insurance benefits/coverage to lower the premiums that will be charged to stay within the original costs allocated in the operating budget (with the difference being paid by employees if they wish).

Regardless of this operator's decision, the operating budget, if affected, must be modified. There are situations in which an operating budget should be legitimately modified; however, an operating budget should never be modified simply to compensate for management inefficiencies.

To illustrate, assume, for example, that a budgeted "total labor" percentage of 25% is realistic and achievable for a specific operation. The operation's managers, however, consistently achieve budgeted sales levels but just as consistently greatly exceed the total labor cost percentage targets established by the operating budget. As this occurs, resulting profits are less than projected.

In the above case, the labor cost portion of the operating budget should not be increased (nor should menu prices be increased!) simply to mask management's inefficiencies in controlling costs in this area. Instead, if the goal of a 25% total labor cost is reasonable and achievable, then the operation's managers must correct the problem producing the higher labor cost percentage.

Properly prepared operating budgets are designed to be achieved, and foodservice operators must do their best to ensure this occurs. There are cases, however, when operating budgets must be modified, or they will lose their ability to assist managers with decision-making responsibilities. The following situations are examples of those that, if unknown and not considered when the original budget was developed, may require operators to consider modifying their existing operating budgets:

✓ Additions or subtractions from offerings that materially affect revenue generation (e.g. reduced or increased operating hours)
✓ The opening of a direct competitor
✓ The closing of a direct competitor
✓ A significant and long-term or permanent increase or decrease in the price of major cost items
✓ Franchisor-mandated operating changes that directly affect (increase) costs
✓ Significant and unanticipated increases in fixed expenses such as mortgage payments (e.g. a loan repayment plan tied to a variable interest rate), insurance, or property taxes
✓ A manager or key employee change that significantly alters the skill level of the facility's operating team
✓ Natural disasters including floods, hurricanes, or severe weather that significantly impact forecasted revenue
✓ Changes in financial statement formats or expense assignment policies
✓ Changes in the investment return expectations of the operation's owners

Operators should have detailed knowledge of their operation, and then they can do a much more effective job comparing their actual performance to their planned (budgeted) performance to identify significant variances. When they do, they can use the cost control strategies and procedures presented in this book to identify problem areas. They can also take the corrective actions needed to ensure that they and their staff can successfully meet all their planned financial goals and objectives.

Technology at Work

The preparation and monitoring of operating budgets in the foodservice industry can be a time-consuming process.

Today, however, there are a variety of software programs available in both PC and Mac formats to help the owners of foodservice businesses develop and monitor their operating budgets.

To review the features of some of these helpful tools, enter "budgeting software for restaurants" in your favorite search engine and view the results.

What Would You Do? 12.2

"Well, it just makes sense to me that we start with how much profit we need to make and then estimate the sales and expenses required to make that profit level. Otherwise, why are we buying the food truck?" said Todd.

Todd was talking to Catriona, his partner in a new joint venture in which Todd and Catriona are sharing the cost of buying a food truck with a menu that will feature English-style Fish and Chips. They had agreed to share the truck's cost and operating profits on a 50–50 basis, and they were now preparing their first year's operating budget.

"I don't know if I agree with that," said Catriona. "It seems to me we need to estimate our realistic sales levels first and then our operating costs to see what profit we're likely to generate in the first year. Just because we want to make a certain amount of profit doesn't mean that we will!"

Assume you were Todd. Why do you think it makes sense to focus on profits first when preparing an operating budget? Assume you were Catriona. Why do you think it is important to focus on estimated sales levels when preparing a first year's operating budget?

Key Terms

Operating budget	Favorable variance	Significant variance
Historical data	Unfavorable variance	Sales mix

Operator's 10-Point Tactics for Success Checklist

Evaluate your need for, and the current status of, each of the following operational tactics. For those tactics you think are important, but not yet in place, develop an action plan for its implementation including who will be responsible for the tactic's completion and the target date by which it should be completed.

Tactic	Don't Agree (Not Done)	Agree (Done)	Agree (Not Done)	If Not Done — Who Is Responsible?	If Not Done — Target Completion Date
1) Operator understands the importance to profits of developing an accurate operating budget.	____	____	____		
2) Operator knows and can identify the unique purposes of each of the three different types of operating budgets.	____	____	____		

Tactic	Don't Agree (Not Done)	Agree (Done)	Agree (Not Done)	If Not Done	
				Who Is Responsible?	Target Completion Date
3) Operator can state the advantages that will accrue when the operation develops and uses well-prepared and detailed operating budgets.	____	____	____		
4) Operator recognizes the importance of assessing prior-period operating results when preparing an operating budget.	____	____	____		
5) Operator recognizes the importance of utilizing realistic assumptions about the next period's operations when preparing an operating budget.	____	____	____		
6) Operator recognizes the importance of understanding the organization's financial objectives when preparing an operating budget.	____	____	____		
7) Operator knows how to create realistic sales forecasts when preparing an operating budget.	____	____	____		
8) Operator knows how to create realistic fixed and variable cost expense forecasts when preparing an operating budget.	____	____	____		
9) Operator has a system in place for identifying significant variances between actual operating results and planned results.	____	____	____		
10) Operator recognizes the importance of corrective action and/or budget modification when actual operating results vary significantly from planned results.	____	____	____		

Glossary

Accountant: An individual skilled in the recording and reporting of financial transactions.

Accounting: The system of recording and summarizing financial transactions and analyzing, and reporting the results, verifying.

Accounting period: The amount of time included in a financial summary or report that should be clearly identified on the financial document. For example, a week, month, or year.

Accuracy in menu laws: Legislation that requires foodservice operations to truthfully, and accurately, represent the quality, quantity, nutritional value, and price of the menu items they sell. Also known as "Truth–in–menu laws" and "Truth–in–dining laws."

À la carte (menu): A menu that lists and prices each menu item separately.

Ambience: The character and atmosphere of a foodservice operation.

Amortization: The practice of spreading an intangible asset's cost over that asset's useful life.

Application: A form, questionnaire, or similar document that employment applicants must complete for the employer. An application may exist in a hard copy, electronic copy, or Internet medium.

Arithmetic average: The simplest and most widely used measure of a mean (average); it is calculated by dividing the sum of a group of numbers by the count of the numbers used in the series. For example, if the reviewers' scores on a review site are 3, 4, 4, 5, and 5, and the sum is 21, the arithmetic mean is divided by five and equals 4.2 (21/5 = 4.2).

Asset (business): Property that is used in the operation of a business including money, real estate, buildings, inventories and equipment.

Auditor: An accounting professional who specializes in studying a foodservice operation's internal controls and analyzing its basic accounting system to ensure all financial information is properly recorded and reported.

Back office accounting system: The accounting software used to maintain a business's accounting records that are not contained in its POS system. This typically includes items such as payroll records, accounts payable (AP) and accounts receivable (AR), taxes due and payable, net profit and loss summaries, and relevant balance sheet entries.

Back-of-house staff: The employees of a foodservice operation whose duties do not routinely put them in direct contact with guests.

Balance sheet: A report that documents the assets, liabilities and net worth (owner's equity) of a foodservice business at a single point in time. Also commonly called the Statement of Financial Position.

Bank reconciliation: A process performed to ensure that a company's financial records are correct. This is done by comparing the company's internally recorded amounts with the amounts shown on its bank statement. Any differences must be justified and, when there are no unexplained differences, accounting personnel can affirm that the bank statement has been reconciled.

Beer: An alcoholic beverage made from malted grain and flavored with hops.

Beginning inventory: The monetary value of all products on hand at the start of an accounting period. Beginning inventory is determined by completing a physical inventory.

Bonding (employee): Protection from loss resulting from a criminal event involving employee dishonesty and theft such as loss of money, securities, and other property.

Bookkeeping: The process of recording a foodservice operation's financial transactions into organized accounts each day.

Bounce rate: The percentage of visitors who viewed only one page before exiting a website.

Brand: The specific ways in which a foodservice operation differentiates itself from its competitors.

Break-even point: The sales point at which total cost and total revenue are equal, and there is no loss or gain for a business.

Bundling: A pricing strategy that combines multiple menu items into a grouping which is then sold at a price lower than that of the bundled items purchased separately.

Business plan: A document that summarizes the operational and financial objectives of a business.

Carafe: A container used for serving wines. A standard carafe holds one standard size (750 ml) bottle of wine. However, carafe size may vary based on operators' service preferences.

Cash bank: The currency and coin holder used to store money received from guests for their cash purchases and to provide change for cash purchases.

Cash receipts journal: An accounting record used to summarize transactions related to cash receipts generated by a foodservice operation.

Classification (expense): Determining where to place an expense on an income statement.

Click-through rate (CTR): The number of clicks an ad or promotion receives divided by the number of times the ad is viewed. The formula used to calculate CTR is: Number of Clicks/Number of Views = CTR. For example, if an operator's promotional ad on a website had 5 clicks with 100 views, the CTR would be 5/100 = 5%.

Collusion: Secret cooperation between two or more individuals that is intended to defraud an operation.

Comp: The foodservice industry term used to indicate food or beverage items that are served free of charge to guests or provided to them at prices lower than the regular menu price. Note: "Comp" is short for "complimentary."

Consumer rationality: The tendency to make buying decisions based on the belief that the decisions will result in a personal benefit.

Content management system (CMS): Computer software used to load content into a digital menu display system.

Contribution margin (CM): The dollar amount remaining after subtracting a menu item's product costs from its selling price.

Controllable (expenses): Costs that can be directly affected by management's actions. Examples include labor and advertising costs.

Controllable income: In a USAR–formatted income statement, the amount of money remaining after a foodservice operation's Prime Cost (Line 14) and Total Other Controllable Expenses (Line 22) are subtracted from Total Sales (Line 4).

Convenience food: Any food item that is partially or fully prepared before it is purchased by a foodservice operation.

Convention and Visitors Bureau (CVB): The local entity responsible for promoting travel and tourism to and within a specifically designated geographic area.

Cost center: A part of a business organization with assignable expenses that generates little or no revenue.

Cost of sales: The total cost of the products used to make the menu items sold by a foodservice operation.

Craft beer: A beer produced in limited quantities and with limited availability.

Curb appeal: The general attractiveness of a foodservice operation when viewed from the outside by a potential customer.

Customer loyalty program: A marketing tool that recognizes and rewards customers who make purchases on a recurring basis. The programs typically award points, discounts or other benefits that increase as the total amount of a repeat customer's purchases increase. Also commonly known as a "Frequent guest program" "Frequent customer program, " or "Frequent dining program."

Cyclical (menu): A menu in which items are offered on a repeating (cyclical) basis.

Depreciation: The allocation of the cost of equipment and other tangible assets based on the projected length of their useful lives.

Digital menu: An integrated system that uses hardware and software to display an operation's menu on an electronic screen; also commonly referred to as a "digital display" menu or "digital menu board."

Dining bubble: A controlled environment used to house foodservice guests while they are dining outdoors.

Domain name: The unique name that appears after "www" in web addresses and after the "@" sign when e-mail addresses are used.

Draft beer An unpasteurized beer product sold in kegs; also known as "tap" beer.

Du jour (menu): A menu featuring items and prices that change daily. Pronounced "duh–zhoor."

EBITDA: Short for "earnings before interest, taxes, depreciation, and amortization." EBITDA is used to track and compare the underlying profitability of a business regardless of its depreciation assumptions or financing choices.

Embezzlement: A crime in which a person or entity intentionally misappropriates assets that were entrusted to the person or entity.

Employee referral: A recruitment approach that occurs when a current foodservice employee refers someone they know to fill an operation's vacant position. Employee referrals are often seen as a good way to find new employees because the referred employee is already known to someone who works in the operation.

Employer of choice: A company whom workers choose to work for when given employment choices. This choice is a conscious decision made when initially joining a company and when deciding to stay with an employer.

Empowerment (employee): An operating philosophy that emphasizes the importance of allowing employees to make independent guest–related decisions and to act on them.

Ending inventory: The monetary value of all products on hand at the end of an accounting period.

Equal Employment Opportunity Commission (EEOC): The entity within the federal government assigned to enforce the provisions of Title VII of the Civil Rights Act of 1964.

Exempt employee: Employees exempted from the provisions of the Fair Labor Standards Act. These workers typically are paid a salary above a certain level and work in an administrative or professional position. The U.S. Department of Labor (DOL) regularly publishes a duties test that can help employers determine who meets this exemption.

Expenses: The price paid to obtain the items required to operate a business. Also referred to as "costs."

Extension (domain name): The combination of characters following the "period" in a web address.

External search (employee): An approach to seeking job applicants that focuses on candidates not currently employed by the organization.

Favorable variance: Better–than–expected performance when actual results are compared to budgeted results.

Fixed asset: An asset such as land, building, furniture and equipment that is purchased for long–term use and is not likely to be converted quickly into cash.

Fixed cost: An expense that remains constant despite increases or decreases in sales volume.

Font size: The size of the characters printed on a page or displayed on a screen. Commonly referred to as "type size."

Foodborne illness: Foodborne illness is caused by consuming contaminated food, beverages, or water infected with a variety of bacteria, parasites, viruses and/or toxins and poisons. These pathogens can be acquired through person–to–person spread, animal contact, the environment and recreational or drinking water.

Food Code (FDA): A model for best practices to ensure the safe handling of food in retail settings.

Food cost percentage: The portion of food revenues spent on food expenses.

4 Ps of Marketing: A way to categorize a business's marketing strategy based on the products sold, places where they are sold, the prices at which they are sold and the promotional efforts used to sell them.

Fraud: Deceitful conduct to manipulate someone into giving up something of value by (a) presenting as true something known by the fraudulent party to be false or (b) by concealing a fact from someone that may have saved him or her from being cheated.

Free-pour: Pouring liquor from a bottle without measuring the poured amount.

Front-of-house staff: The employees of a foodservice operation whose duties routinely put them in direct contact with guests.

F-shaped scanning pattern: A text scanning pattern characterized by fixations concentrated at the top and the left side of a page. Website viewers typically first read in a horizontal movement across the upper part of the content area. Notes: These areas initially form the F's top bar, and viewers then scan down. The "F" in F-shaped scanning pattern is short for "Fast" and the method is also commonly referred to as the "F–based reading pattern."

Furniture, fixtures, and equipment (FF&E): Movable furniture, fixtures, or other equipment that have no permanent connection to a building's structure.

General ledger: The main or primary accounting record of a business.

Ghost kitchen: A foodservice operation that provides no on–site dine–in services and that prepares all of its menu items only for pick up or delivery to guests. Also known as a delivery–only restaurant, shadow kitchen, cloud kitchen, or virtual kitchen.

Google search: Also known simply as "Google," this is a method of finding information on the Internet. Note: Google conducts over 92% of all Internet searches and is the most visited website in the world.

Green practices: Those activities that lead to more environmentally friendly and ecologically responsible business decisions. Also known as "eco–friendly" or "earth–friendly" practices.

Guest check average: The average (mean) amount of money spent per guest (or table) during a specific time. Also referred to as "check average" or "ticket average."

Health inspector: A professional responsible for ensuring compliance of sanitation codes required of businesses that are open to the public. This compliance is achieved through facility inspections that may be unannounced or occur as a result of a complaint.

Historical data: Information about an operation's past financial performance including revenue generated, expenses incurred, and profits (or losses) realized.

Income before income taxes: The amount of money remaining after an operation's interest expense is subtracted from the amount of its restaurant operating income. Also, a business's profit before paying any income taxes due on the profits.

Income statement: Formally known as "The Statement of Income and Expense," a report summarizing a foodservice operation's profitability including details regarding revenue, expenses and profit (or loss) incurred during a specific accounting period. Also commonly called the Profit and Loss (P&L) statement.

Interest expense: The cost of borrowing money.

Internal search (employee): A promote–from–within recruitment approach used to identify qualified job applicants.

IP address: A unique address that identifies a device on the Internet or a local network. IP stands for "Internet Protocol," a set of rules governing the format of data sent via the Internet.

Jigger: A small cup–like bar device used to measure predetermined quantities of alcoholic beverages. These items are usually marked in ounce and portions of an ounce. Examples: 1 ounce or 1.5 ounces.

Job description: A statement that outlines the specifics of a particular job or position within a foodservice operation. It provides details about the responsibilities and conditions of the job.

Job specification: A listing of the personal characteristics and skills needed to perform the tasks contained in a job description.

Keywords: The words and phrases that users type into search engines to find information about a particular topic. Also known as "SEO keywords," "Key phrases," and "Search queries."

Labor cost percentage: A ratio of overall labor costs incurred relative to total revenue generated.

Labor dollars per guest served: The dollar amount of labor expense spent to serve each of an operation's guests.

Law of demand: The law of demand holds that the demand level (number of units sold) for a product or service declines as its price rises and increases as the price declines.

Link: An easy way to navigate from one webpage to another webpage, or from one section of a webpage to another section of the same webpage.

Local link: Website links created to show that other entities with relevance in the local area trust or endorse a foodservice operation.

Management: The process of planning, organizing, directing, controlling, and evaluating the financial, physical, and human resources of an organization to reach its goals.

Managerial accounting: The accounting specialty that uses historical and estimated financial information to help foodservice operators plan the future.

Mandatory benefit: An employee benefit that must, by law, be paid by an employer.

Market: The group of all people who could be customers of a business.

Marketing: The varied activities and methods used to communicate a business's products and service offerings to its current and potential guests.

Marketing calendar: A schedule detailing all marketing activities planned for the time period covered by the marketing plan.

Marketing channel: A method or form of delivering a marketing message to an operation's current and potential guests. Also commonly known as a "communication channel." Examples include print, broadcast, and web–based communication methods.

Marketing message: How an operation communicates to its customers and highlights the value of the products and services that are offered to its target markets.

Marketing mix: The specific ways a business utilizes the 4 Ps of marketing to communicate with its potential customers.

Marketing plan: A written plan that details the marketing efforts of a foodservice operation for a specific time (usually annually).

Marketing return on investment (ROI): The amount of benefit gained from a marketing activity compared to the cost of undertaking and implementing the activity.

Marketing strategies: The broad and long–term marketing goals a foodservice operation wants to achieve.

Marketing tactics: The specific steps and actions undertaken to achieve marketing strategies (goals).

Menu: A French term meaning "detailed list." In common usage it refers to (i) a foodservice operation's available food and beverage products and (ii) how these items are made known to guests.

Menu copy: The words and phrases used to name and describe items listed on a foodservice menu.

Menu engineering: A system used to evaluate menu pricing and design by categorizing each menu item into one of four categories based on profitability and popularity.

Moment of truth: A point of interaction between a guest and a foodservice operation that enables the guest to form an impression about the business.

Mystery shopper: A person posing as a foodservice guest who observes and experiences an operation's products and services during a visit and who then reports findings to the operation's owner. Also referred to as a "secret shopper."

NAP: An abbreviation for an operation's name, address, and phone number.

Navigation (website): The process of moving from one content section of a website to another section of the same website.

Negligent hiring: Failure of an employer to exercise reasonable care when selecting employees.

Negligent retention: Retaining an employee after the employer becomes aware of an employee's unsuitability for a job by failing to act on that knowledge.

Noncash expense: Expenses recorded on the income statement that do not involve an actual cash transaction. Examples include depreciation and amortization expenses where an income statement charge reduces operating income without a cash payment.

Noncontrollable expenses: Costs which, in the short run, cannot be avoided or altered by management decisions. Examples include lease payments and depreciation.

Nonprofit sector (foodservice): Foodservice in an organization where generating food and beverage profits is not the organization's primary purpose. Also referred to as the "noncommercial" sector.

Nutrient: A chemical compound such as protein, fat, carbohydrate, vitamin, and/or minerals contained in foods that are used by the body to help it function and grow.

Objective test: An assessment tool such as a multiple choice or true/false test whose questions have one correct answer and yield a reduced need for trainers to interpret trainees' responses.

Occupancy costs: Costs related to occupying a space including rent, real estate taxes, personal property taxes, and insurance on a building and its contents.

Off-premise menu: A menu listing items available for pickup or delivery; also commonly referred to as a takeaway, take-out, or carry-out menu.

Open check: A guest check that initially authorizes product issues from the kitchen or bar for which collection of the amount due has not been made (the account is not closed). Therefore, the amount due has not been added to the operation's sales total.

Open-door communication (policy): A policy that indicates supervisors or managers are willing to discuss an employee's questions, complaints, suggestions, and concerns at convenient times.

Operating budget: An estimate of the income and expenses of a business over a defined accounting period. Also referred to as a "financial plan."

Organic (food): the product of a farming system which avoids the use of synthetic fertilizers, pesticides, growth regulators, and livestock feed additives. The U.S. Department of Agriculture (USDA) requires that food products sold, labeled, or represented as organic in the United States must have at least 95% certified organic content.

Other Controllable Expenses: Expenses that a foodservice operator can influence with increases or decreases based on business decisions. Examples include marketing costs and utility costs.

Over (cash bank): A cashier's bank that contains more than the amount of cash, credit, or debit card, or other authorized payment form based on the number of menu items sold.

Paid social advertising: The use of ads on social media sites for which an operator must pay a fee.

Payroll: The term commonly used to indicate the amount spent for labor in a foodservice operation. Used for example in "Last month our total payroll was $28000."

PDF: Short for "Portable Document Format:" a document format that allows this information to be shown clearly regardless of the software, hardware, or operating system used by the viewer.

Percentage variance: The percentage difference between an operation's actual and estimated income or expense.

Person-in-charge (PIC): The owner of the business, or a designated person–such as a chef, kitchen manager, or employee who is present at the work site and has direct authority and supervision over employees who engage in the safe storage, preparation, display, and service of food.

Physical inventory: An inventory control system tool in which an actual (physical) count and valuation of all product inventory on hand is taken at the close of an accounting period.

Point-of-sale (POS) system: An electronic system that records foodservice customer purchases and payments and other operational data.

Portion cost: The product cost required to produce one serving of a menu item.

Positive cash flow: The condition that exists if the amount of a business's cash that is acquired exceeds the amount of cash spent by the business. Stated differently, positive cash flow means more cash is coming into a business than is going out of the business, and this is essential for the business to sustain long–term growth.

Preopening marketing expense: A nonrecurring promotional or advertising cost incurred before a new operation opens. Also referred to as a "start–up" marketing cost.

Pretest/post-test evaluation: A before and after assessment used to measure whether the expected changes took place in a trainee's knowledge, skill level, and/or attitude after the completion of training.

Price (Noun): A measure of the value given up (exchanged) by a buyer and a seller in a business transaction. For example: "The price of the Mushroom Swiss Burger combo meal is $9.95."

Price (Verb): To establish the value to be given up (exchanged) by a buyer and a seller in a business transaction. For example: "We need to price the Mushroom Swiss Burger combo meal."

Prime (beef): The highest quality grade given to beef graded by the USDA. Prime beef is produced from young, well–fed beef cattle, and it has abundant marbling (the amount of fat interspersed with lean meat).

Prime cost: An operation's cost of sales plus its total labor costs.

"Prime real estate (menus"): A phrase used to define the areas on a menu that are most visible to guests, and which should contain the items menu planners most want to sell (those that are most popular and profitable)."

Productivity (worker): The amount of work performed by an employee within a fixed amount of time.

Profit: The dollars that remain after all a business's expenses have been paid. Also referred to as "Net income."

Profit center: A part of a business organization with assignable revenues and expenses. Also referred to as a "revenue center."

Profit margin: The amount by which revenue in a foodservice operation exceeds its operating costs.

Proof (alcoholic beverage): A measure of the alcohol content of an alcoholic beverage. In the United States, alcohol proof is defined as twice the percentage of "alcohol content in percent by volume (ABV)."

Proprietary website: A website in which the foodservice operator controls all the website's content and can readily make changes to it.

Public record: Documents or other information that are not considered confidential, that generally pertain to the conduct of a government entity, and that can be viewed by the general population.

Quick-change artist: An individual who attempts to confuse a cashier into giving excessive change.

Quick service restaurants (QSRs): Foodservice operations that typically have limited menus that often include a counter at which customers can order and pick up their food. Most QSRs also have one or more drive–thru lanes that allow customers to purchase menu items without leaving their vehicles. QSRs may also offer off–site delivery services for their menu items. Menu prices in QSRs are normally lower when compared to some other restaurant types.

Ratio: An expression of the relationship between two numbers. Ratios are computed by dividing one number by the other number.

Receivable revenue: The expected sales income when all food and beverage orders have been properly documented.

Recipe conversion factor (RCF): A mathematical formula that yields a number (factor) operators can use to convert a standardized recipe with a known yield to the same recipe that produces a desired yield.

Registered Dietitian (RD): A health professional who has special training in diet and nutrition. RDs offer advice on nutrition and healthy eating habits to help people improve their health and well–being.

Registration (domain name): The act of reserving a domain's name on the Internet for a fixed time (usually one year). Registration is necessary because domain names cannot typically be purchased permanently.

Restaurant operating income: All an operation's revenue minus all of its controllable and noncontrollable expenses.

Restaurant row: A street or region well–known for having multiple foodservice operations a close distance from each other.

Return on investment (ROI): A measure of the ability of an investment to generate income.

Revenue: The term indicating the dollars taken in by a business within a defined time period. Also referred to as "sales."

Revenue metric: A standard for measuring or evaluating data based on its dollar amount or change in dollar amount. Also known as a "financial metric."

Salaried employee: An employee who regularly receives a predetermined amount of compensation each pay period on a weekly or less frequent basis. The predetermined amount paid is not reduced because of variations in the quality or quantity of the employee's work.

Sales mix: The series of individual guest purchasing decisions that result in a specific overall food or beverage cost percentage.

Sales per labor hour: The dollar value of sales generated for each labor hour used.

Scaling (recipe): The process of adjusting the yield of a standardized recipe.

Search engine optimization (SEO): The process of improving and modifying an operation's website with the purpose of increasing its visibility when users search for products and/or services related to the operation.

Search engine results page (SERP): The webpage of a search engine such as Google, Bing, or Yahoo that shows a user when a user types in a search query.

Service industry: A business segment that primarily provides services to its customers.

Service recovery: The actions taken to correct the results of a poor customer service experience (moment of truth) and to regain customer loyalty.

Short (cash bank): A cashier's bank that contains less than the amount of cash, credit, or debit card, or other authorized payment form based on the number of menu items sold.

Significant variance: A variance that requires immediate management attention.

Social media: A collective term for websites and applications that enables users to create and share information and content or to participate in social networking.

Spirits (beverages): Alcoholic beverages produced by the distillation of fermented grains, fruits, vegetables, and/or sugar.

Standardized recipe: The instructions needed to consistently prepare a specified quantity of food or drink at an expected quality level.

Standard operating procedure (SOP): The term used to describe how something should be done under normal business conditions.

Statement of Cash Flows (SCF): A report providing information about all cash inflows a company receives from its on–going operations and external investment sources. It also includes all cash outflows paying for business activities and investments during a defined accounting period.

Table d'hôte (menu): A menu with a pre-selected number of menu items offered for one fixed price. Pronounced "tah–buhlz–doht." Also known as a "prix fixe" (pronounced prē–'fēks) menu.

Target market: The group of people with one or more shared characteristics that an operation has identified as the *most likely* customers for its products and services.

Third-party delivery: A smart phone or computer application that creates a marketplace that customers can search to browse restaurant menus, place orders, and have them delivered to a location of the customer's choosing. In nearly all cases the guest orders are delivered by independent contractors who have been retained by the company operating the third–party delivery app.

Ticket time: The amount of time required to fill a guest's order.

Timing (expense): Determining when to place an expense on an income statement.

Topline revenue: Sales or revenue shown on the top of the income statement of a business.

Total labor: The cost of the management, staff, and employee benefits expense required to operate a business.

Total sales: The sum of food sales and alcoholic beverage sales generated in a foodservice operation.

Traffic tracking (website): Information that tells website operators, among other things, the number of site visitors their websites have had, how the site was found, how long visitors browsed different pages on the website, and how frequently visitors were converted into customers.

Training: The process of developing a staff member's knowledge, skills, and attitudes necessary to perform required tasks.

Training handbook: A hard copy or electronic manual containing the training plan and associated training lessons for a foodservice operation's complete training program.

Training lesson: Information about a single session of a training plan. It contains one or more training objectives and indicates the content and method(s) to enable trainees to master the content.

Training plan: A description of the overview and sequence of a complete training program.

Train-the-trainer (program): A training framework that turns employees into subject matter experts who can teach others.

28-day accounting period: An accounting period that is four weeks (28 days) in length instead of a calendar month that has between 28 and 31 days. There are 13 four–week periods instead of 12 monthly periods when using this system.

Umami: "Essence of deliciousness" in Japanese, and its taste is often described as the meaty, savory deliciousness that deepens flavor.

Unfavorable variance: Worse–than–expected performance when actual results are compared to budgeted results.

Uniform System of Accounts for Restaurants (USAR): A recommended and standardized (uniform) set of accounting procedures used for categorizing and reporting restaurant revenue and expenses.

Up-selling: The process of increasing the guest check average through the effective use of on–site selling techniques; also referred to as "suggestive selling."

User experience: The feelings resulting from a user's interactions with a digital product.

User-generated content (UGC) site: A website in which content including images, videos, text, and audio have been posted online by the site's visitors.

Value: The amount paid for a product or service compared to the buyer's view of what they receive in return.

Value proposition: A statement that clearly identifies the benefits an operation's products and services will deliver to its guests.

Variable cost: An expense that generally increases as sales volume increases and decreases as sales volume decreases.

Variance: The difference between an operation's actual and its estimated income or expense.

Voluntary benefit: An employee benefit paid at the discretion of an employer in efforts to attract and keep the best possible employees.

Wagyu beef: Beef from a Japanese breed of cattle that is highly prized for its marbling and flavor. In the Japanese language, "Wa" means Japanese, and "gyu" means cow.

Walk (bill): Guest theft that occurs when a guest consumes products that were ordered and leaves without paying the bill. Also referred to as a "skip."

Web host: An organization that provides space for holding and viewing the contents of a website. All websites and e–mail systems must be hosted to connect an operator's proprietary website content to the rest of the Internet.

Weighted contribution margin: The contribution margin provided by all menu items divided by the total number of items sold. Weighted contribution margin is calculated as: Total Contribution Margin of All Items Sold/Total Number of Items Sold = Weighted contribution margin.

Wine: An alcoholic beverage made from fruit; most typically grapes.

Wine list: A foodservice operation's wine menu.

Index